CONTENTS

COURSE CONTENT AND ASSESSMENT

SYLLABUS

The main purpose of this book is to help you improve your chances of success within the National 5 Engineering Science course, and to act as a supplement to your learning in school. Regularly studying this book, and using it to consolidate your learning in the classroom, will go a long way towards success in this course.

The course is broken up into three main units, and each chapter within this book covers the content you need to know and understand. These units are:

- Engineering contexts and challenges
- Electronics and control
- Mechanisms and structures.

Engineering contexts and challenges

This part of the course gives a broad context of engineering. While completing this, you will develop an understanding of engineering concepts by exploring and analysing a range of engineered objects, engineering problems and solutions. You will investigate existing and emerging technologies and engineering challenges, considering any implications that will arise from their solutions.

Electronics and control

Within this part of the course, you will study several key concepts used within electronic control systems, including analogue, digital and programmable electronics. This can be done through simulation or practical projects, and you will explore several different engineering problems and solutions in a range of contexts.

Mechanisms and structures

This unit is designed to help develop your understanding of mechanisms and structures. This can be done through simulation or practical projects. You will learn to investigate tasks in a range of contexts, and to discover problems, and through evaluation you will be able to make suggestions for improvement.

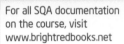

ONLINE

For all SQA documentation on the course, visit www.brightredbooks.net

ASSESSMENT

To gain the full course award, you must pass the course assessment. This is broken into two separate parts:

- **Part 1** – assignment (50 marks)
- **Part 2** – question paper (110 marks)

The marks of these two components will be combined and your award will be based on your total score. This means your overall assessment will be marked from a total of 160 marks.

You will gain a lot of knowledge and skills throughout this course. The purpose of the assignment is to assess these in a practical way, developing a solution to a challenging engineering problem. This is 31% of your overall grade, so it is worthwhile putting in the effort to ensure it is completed to the best of your ability.

This part of the assessment is closed book, which means you will not have access to your notes, the internet or any such resources.

The assessment of this project will be broken up into separate sections:

contd

BrightRED Study Guide

Curriculum for Excellence

N5

ENGINEERING SCIENCE

Paul MacBeath

First published in 2016 by:
Bright Red Publishing Ltd
1 Torphichen Street
Edinburgh
EH3 8HX

A CIP record for this book is available from the British Library.

ISBN 978-1-906736-69-9

With thanks to:
Ivor Normand (editorial) and PDQ Digital Media Solutions (layout)
Cover design and series book design by Caleb Rutherford – e i d e t i c.

Acknowledgements
Every effort has been made to seek all copyright-holders. If any have been overlooked, then Bright Red Publishing will be delighted to make the necessary arrangements.

Permission has been sought from all relevant copyright holders and Bright Red Publishing are grateful for the use of the following:

RyanKing999/iStock.com (p 6); Stefan Krause License (LAL)[1] (p 10); US EPA (Public domain) (p 12); Caleb Rutherford (e i d e t i c) (p 13); Capgros/freeimages.com (p 17); raining girl (CC BY 3.0)[2] (p 18); Energy Information Administration, Geothermal Energy in the Western United States and Hawaii: Resources and Projected Electricity Generation Supplies, DOE/EIA-0544 (Washington, DC, September 1991) (Public domain) (p 18); Svjo (CC BY-SA 4.0)[3] (p 25); Stu49/Shutterstock.com (p 28); Timawe/iStock.com (p 29); Shur23/iStock.com (p 32); Bisgaard (CC BY-SA 3.0)[4] (p 33); SparkFun Electronics (CC BY 2.0)[5] (pp 43 & 48); Ravi Kotecha (CC BY-SA 4.0)[3] (p 43); © Oorka | Dreamstime.com – Traffic Lights Photo (p 45); WARDJet (CC BY-SA 3.0)[4] (p 47); Arduino® is a registered trademark of Arduino LLC (pp 48–55); Arduino (CC BY-SA 3.0)[4] (p 48); Arduino (CC BY-SA 3.0)[4] and Fritzing UG (p 49); Arduino (CC BY-SA 3.0)[4] (p 50); Arduino (CC BY-SA 3.0)[4] and Fritzing UG (p 51); Patrick McGarvey (CC BY-ND 2.0)[6] (p 51); Jjmontero9 (CC BY-SA 3.0)[4] (p 54); Arduino (CC BY-SA 3.0)[4] and Fritzing UG (p 54); (p 55); Patrick McGarvey (CC BY-ND 2.0)[6] (p 55); Kimsaldo (CC BY-SA 4.0)[3] (p 59); Pastorius (CC BY-SA 3.0)[4] (p 63); Patrick (CC BY-SA 2.0)[7] (p 63); Myfuture.com (CC BY-ND 2.0)[6] (p 78); Andyminicooper (CC BY-SA 3.0)[4] (p 82); Supermac1961 (CC BY 2.0)[5] (p 84); Senior Airman Max Rechel (Public domain) (p 84); Pierre-Yves Beaudouin (CC BY-SA 4.0)[3] (p 84); NAVFAC (CC BY 2.0)[5] (p 84); NBC News (CC BY 2.0)[5] (p 87); Benson Kua (CC BY-SA 2.0)[7] (p 87); Jarek Tuszyński (CC BY 4.0)[8] (p 87); Dave Haygarth (CC BY 2.0)[5] (p 88); Konstantin Tyurpeko (CC BY-SA 4.0)[3] (p 88); Divulgação Petrobras/Abr (CC BY 3.0 BR)[9] (p 89); CollegeDegrees360 (CC BY-SA 2.0)[7] (p 90); Images licensed by Ingram Image (pp 6, 7, 9, 11–17, 19–22, 30, 35–36, 42, 45–46, 55, 57–58, 60–63, 67–68, 74, 78, 80–81, 83 & 89).

Printed and bound in the UK.

Area	Range of Marks
Analysing the problem (specification, system diagram, sub-system diagram and a description of it)	4–8
Designing a possible solution (flowchart, design sketches of mechanisms and structures with calculations)	8–12
Building a solution (producing code if necessary and modelling/simulating solutions to all mechanical/electronic parts)	8–12
Testing the solution (test plans and then testing of solution)	8–14
Reproducing and evaluation of the solution (photos/printouts of solution, reflection of development process, evaluation of solution)	8–14

Question paper

The purpose of the question paper is to assess your breadth of knowledge from across all the units. Here you will display your depth of understanding and have the opportunity to apply this knowledge and understanding to answer appropriately challenging questions.

The question paper will have 110 marks, which is 69% of the total mark.

Approximately 20–30% of this will come from the 'Engineering contexts and challenges' section of the course. Approximately 30–40% of the marks will be questions related to the 'Electronics and control' part of the course, and approximately 30–40% of the marks will come from the 'Mechanisms and structures' part of the course.

The exam will have two different sections.

- **Section 1** will be out of 20 marks, and will consist of short-answer questions.
- **Section 2** will be out of 90 marks, and will be made up from longer structured questions that will combine several areas of the course.

If you are unsure of an answer during the exam, then miss it out and come back to it at the end instead of wasting time thinking of a possible solution, but **always** come back to it. If you are still not sure, then use the data booklet to help, and at least have a guess. If you guess, you may pick up some marks. If you leave it blank, it is guaranteed you will lose all marks on this question.

 ## THINGS TO DO AND THINK ABOUT

When answering calculations on this course, make sure you are using significant figures within your final answers. In a number, all figures are significant, except zeros at the front (to show decimal points) or at the end. For example, 1234 has four significant figures. 0·012 has two significant figures, as does 1200.

The number of significant figures you use in your final answers should be equivalent to those used within the question.

DON'T FORGET

Make sure you read through the brief and marking instructions to ensure you are doing **everything** needed to succeed. Use a highlighter if you need to, so you can highlight important parts.

DON'T FORGET

This will **not** be 'open book', but you will be provided with a data booklet that contains a lot of important calculations and data you will need. Use this to help you!

ONLINE

Follow the link at www.brightredbooks.net to the data booklet. If your teacher hasn't already given you a copy, it is worthwhile printing out to help you answer your coursework.

ONLINE

Follow the link at www.brightredbooks.net to the official SQA past papers and answer schemes. It is highly recommended that throughout this course you visit these and try to answer the questions for the topic you are studying. One of the best ways to prepare for an exam is to answer exam questions!

ENGINEERING SYSTEMS

WHAT IS AN ENGINEER?

Engineering is one of the most important jobs in the world today, as it relates to everything in your life. An engineer is behind everything – from the house you live in, to the clothes you wear, to the deodorant you put on this morning. Engineers shape the world we live in. They create, design, test and improve almost every process or product you know.

Do you wear make-up? Or wear perfume or aftershave? A chemical engineer would have been involved in the creation and testing of it. Do you own a smartphone? An electronics engineer would have played a huge role in its conception and manufacture. Do you travel to school by bus or car? Mechanical engineers have worked in a team to help design these, and civil engineers would have helped to create the roads they drive on.

Engineering plays a big part in a vast array of businesses throughout Scotland, the UK and the world. According to several reputable sources, such as the Engineering Development Trust, Scotland is home to the 'engineering capital of Europe', Aberdeen. This has been because of the oil and renewable energy industries; but engineering is not just working offshore on an oil rig. Engineers work in lots of different settings – in offices, laboratories, recording studios, hospitals, underground and at sea. Becoming an engineer can also lead you into many industries you may not have thought about, such as medicine, food, fashion, sports, space exploration, the gaming industry, transport, construction and much more. With engineering, you can follow your interests – if sport is your thing, you can work as an engineer improving the performance of new running shoes or sports clothing, or help to design a new football for the World Cup. If you want to make a difference to people's lives, you can help to develop artificial limbs or help to rebuild a community following a natural disaster.

VIDEO LINK

Head to
www.brightredbooks.net
to watch the clips exploring
engineering further.

TYPES OF ENGINEER

Although there is a huge array of paths down which an engineering career could lead you, there are seven main branches covered within this course: **environmental**, **civil**, **structural**, **mechanical**, **chemical**, **electrical** and **electronic**.

Environmental engineer

This branch concerns itself with protecting life – whether this is people, animals or plants – from adverse environmental effects such as pollution. Environmental engineers work to improve recycling, water and air quality, waste disposal and public health.

Civil engineer

Civil engineers deal with infrastructure. They plan, design and oversee the construction and maintenance of building structures and facilities, such as roads, railways, bridges, dams, irrigation projects, power plants, and water and sewerage systems.

Structural engineer

Structural engineers analyse, design, plan and research structures, ensuring that a structure is safe and can support the needed weight. They are trained to understand and calculate the stability, strength and rigidity of a structure, ensuring that it will not collapse under certain loads, forces or conditions.

Mechanical engineer

Mechanical engineers concern themselves with anything that moves. They design and develop mechanical devices to complete specific jobs.

Chemical engineer

Chemical engineers deal with the chemical properties of materials and how these can be changed or altered for specific jobs. This could be from coating a metal with something to make it more water- and rustproof, to making plastic from oil.

contd

Electrical engineer

This branch of engineering can be thought of to include not only electronics but also power generation and distribution, motors and electromechanical devices. It concerns itself with power generation, transmission, utilisation and measurement.

Electronics engineer

Although the name is similar to electrical engineering, it is different. Electronic engineering is about automatic control and the implementation of it. It encompasses analogue and digital circuits as well as computer programming.

You will discover that, although there are distinctly different branches of engineering, these disciplines can overlap and encompass several different specialisms. These engineers also frequently come together to work in teams for different engineering projects.

Consider the Falkirk Wheel. This is a rotating boat lift that was designed to connect the Forth and Clyde Canal and the Union Canal, which is 24 metres above. Within the designing and creating of this, many engineers were involved, such as mechanical engineers who had to create the mechanisms inside to allow the Wheel to move safely and efficiently, and structural engineers to make sure the actual structure was strong enough to lift the boats safely with people and cargo inside. Chemical engineers were involved to ensure the materials were strong enough to do the job, and that they wouldn't be eroded or damaged by being in frequent contact with water. Electrical engineers had to make sure there was enough power to actually do this lifting, as well as designing the lighting systems and other electrical aspects within the project. Civil engineers were also involved in designing and planning how the waterways would link, ensuring that the ground surrounding the water would be capable of building on, as well as planning and designing the road systems going to and coming from the attraction.

IMPACTS OF ENGINEERING

Within any engineering challenges or solutions, we also have to consider the impacts that these will have on the population and the surrounding areas, which can be positive as well as negative.

We have to consider **social impacts:**
— Are there increased employment or training opportunities created through this engineering project?
— Will there be improved infrastructure because of it?
— Will there be traffic disruption because of its creation?
— Will there be disturbances because of noise?

We also have to consider **environmental impacts:**
— Could this provide habitats for wildlife?
— Will there be a risk of damaging animal habitats or ecosystems?
— Will there be a loss of green belt? Will nature be destroyed/damaged because of it?
— Will there be a risk of danger to animals because of this engineering solution?
— Will there be more demand on water or power services?

We also have to consider **economic impacts:**
— Will this engineering solution bring money into the local area through tourism or other means?
— Will this attract other companies to invest in the area?
— Will this employ more people, meaning more money spent in the local area?

EMERGING TECHNOLOGIES

Another thing that engineers have to consider when coming up with a solution is emerging and new technologies. Is there a new or emerging material that would be more suitable for the task? Is there a new process for completing the task that will be better suited for the job? Has something been created that makes it more economical or efficient? If it is powered, could it be run on a different type of fuel that is more environmentally friendly?

THINGS TO DO AND THINK ABOUT

Consider a current engineering solution. Research it and find out what types of engineers contributed to its design and construction. What impacts did this solution have on the local area? Find positive and negative examples of social, environmental and economic impacts it has had. Are there any examples of emerging technology used within the solution? It would be useful to bring your findings into class to discuss them.

ONLINE TEST

Head to www.brightredbooks.net to test yourself on engineering systems.

THE SYSTEMS APPROACH

THE UNIVERSAL SYSTEM DIAGRAM

As you know, engineers are in the business of solving problems. Whether it's planning a new bridge to cross a river, designing the electronics on a new communications system, or creating a sub-sea ROV, engineers will be given a problem that they have to create the solution for. The first step to solving this problem is creating a **universal system diagram** using the systems approach. Most things can be analysed using this method.

When creating a universal system diagram for a particular problem, ask yourself the following questions:

— What **inputs** need to go into the system to make it work?
— What is the name of the **process** you are trying to describe? We do not need to know how it is done, so the process is usually put inside a 'black box'.
— What **outputs** will come out of the system? (Remember, some outputs will be unwanted!)

It might help you to understand the idea of a system if you think about your school bus. The driver only needs to know how to drive and to ensure there is fuel in it. How it all works can be left to the mechanics in the garage, and the output will be the bus driving.

TYPES OF ENERGY

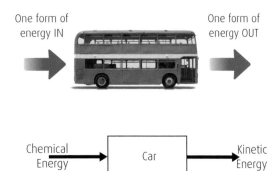

To get a clearer picture, the universal system diagram is frequently shown using the energy that is inputted and outputted. Energy is needed to make things work, and it can be converted from one form into another by a suitable process.

Below are the main forms of energy:
- Sound
- Heat
- Light
- Electrical
- Chemical
- Nuclear
- Field (magnetic)
- Kinetic (movement)
- Potential (stored).

Take our example of the bus. Its input of fuel will be a form of chemical energy. Its output is movement, which is kinetic energy.

SUB-SYSTEM DIAGRAMS

The **universal system diagram** is a very basic way for working out how technology works, but to get a greater understanding of how the system works we have to break it down into more detail. To do this, we draw something called a **sub-system diagram**.

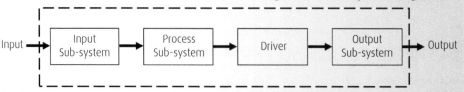

A sub-system diagram is used to show the internal features of the system. Each box, known as a sub-system, can be thought of as a system within a system, and it has its own input and

contd

output. The dashed box around these sub-systems is known as the **system boundary**, and this surrounds the area of interest. Anything inside this box is part of the overall system. Anything outside this, such as the input and output, are part of the 'outside world'.

By using a sub-system diagram, it shows more detail of how a system works. For example, a heater:

By creating a universal system diagram for this, we can discover what goes in and comes out, but we do not gather much information on how the system actually works. But, by breaking it down into the sub-system diagram, it is clearer how the technology works.

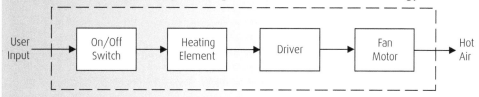

OPEN-LOOP CONTROL

Open-loop control systems have no form of feedback. This is the simplest and most common level of control, as it is cheap to install and simple to operate. However, although open-loop control has many uses, its basic weakness lies in its inability to adjust the output to suit the requirements.

CLOSED-LOOP CONTROL

Sub-system diagrams can also have closed-loop control. This is the most sophisticated form of control. In a closed-loop system, the output is constantly monitored to ensure that the reference value (one that is set by the user) matches the actual value. If there is any difference between them, the system will change things to reduce the output error to zero.

A closed-loop sub-system diagram for the heater is shown below.

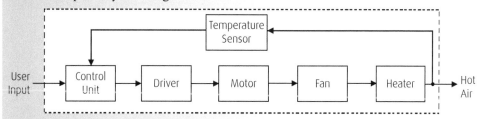

All closed-loop control systems include a feedback loop with some form of sensing sub-system that will feed information back to the control sub-system. The control sub-system will process this feedback signal and then decide whether it needs to alter things to ensure the desired output.

THINGS TO DO AND THINK ABOUT

Look around your house and find open-loop and closed-loop controlled items. Draw the universal system diagram and sub-system diagram for them.

VIDEO LINK

Watch the video at www.brightredbooks.net to further clarify the difference between open-loop and closed-loop systems.

DON'T FORGET

A driver is always needed in a sub-system diagram, as it amplifies the current and power to the output, allowing it to run.

DON'T FORGET

Always make sure you draw proper arrows when you are creating system diagrams – you could lose marks in the final exam if you do not!

DON'T FORGET

Closed-loop control always has a feedback loop, open-loop control does not.

ONLINE TEST

Head to www.brightredbooks.net to test yourself on system diagrams.

ENERGY: FOSSIL FUELS AND RENEWABLES

WHAT IS ENERGY?

Energy is all around us, although it cannot always be seen or touched. A ball is given energy when we kick it. Cars are given energy through petrol or diesel; and we get the energy to walk, talk and run by eating food. Without energy, nothing can happen.

Energy comes in many forms and can be converted from one form to another, but it **cannot** be destroyed.

WHERE DOES ENERGY COME FROM?

All energy ultimately comes from the sun, and it can be created in one of three ways:

- Nuclear fuels
- Renewable energy
- Biosphere fuels.

NUCLEAR FUELS

We mine for uranium which contains radioactive atoms created from the 'Big Bang'. The nuclear-energy process is highly controversial, however, as it creates extremely hazardous waste, and there would be a catastrophe if an accident occurred. It is still used throughout the world though, as many believe the positives can outweigh the potential negatives through the huge amount of energy it can create from a very small amount of ore. To create this energy, uranium atoms are split at nuclear power stations in an effort to speed up the decaying process of the material. From this, we then capture the energy it is creating. This is known as nuclear fission. The heat produced by this process can then be used to create steam, which in turn is used to power a generator to produce even more energy in the form of electricity.

RENEWABLE ENERGY

ONLINE

Find more information on renewable energies by going to www.brightredbooks.net

We can capture the sun's energy directly through the use of solar panels, and indirectly through wind turbines and wave and tidal power stations. This type of fuel can be used over and over again as long as the sun shines and the Earth is still alive. With recent technological advancements, this form of energy is becoming more widely used, and Scotland is quickly becoming one of the world leaders in this field.

Renewable energy can be harnessed domestically on a small scale, but also on a larger industrial scale. Wind, wave, tidal, hydro-electric and solar energy are all commonly used, and each has its own advantages and disadvantages.

Type of energy	How it is harnessed	Advantages	Disadvantages
Wind power	Wind turbines	Free energy No harmful gases produced	Weather is unpredictable May spoil views Dependent on the wind: no wind, no electricity created
Wave power	Wave-energy converter	Free energy No harmful gases produced	Only suitable at certain locations May block fishing/travel routes Will affect marine-life ecosystem
Tidal power	Tidal-energy generator	Free energy No harmful gases produced	Only suitable at certain locations Will affect marine-life ecosystem Tides only happen twice a day, so will only produce energy then
Hydro-electric	Dam and hydraulic turbine	Free energy No harmful gases produced	Only suitable at certain locations Will affect marine-life ecosystem
Solar	Solar panels, solar collectors or photocells	Free energy No harmful gases produced	Weather is unpredictable Dependent on the sun: no sun, no energy harnessed

FOSSIL FUELS

Plants make chemical compounds by a process called photosynthesis. By doing this, the plants store a lot of energy that can then be made available to the Earth's living system. After many millions of years, they will become fossilised and become what is known as a fossil fuel. They come in a very concentrated form that holds a huge amount of stored energy which is easily converted into a readily usable energy. These come in three main forms:

- Coal
- Oil
- Gas.

Coal

Coal, as well as peat, was formed by the decomposition of large plants in a watery environment millions of years ago. The pressure of sand and rock over many years compressed this decomposed material, squeezing out any form of liquid from within to create a rock-like material.

Oil

Oil was created in shallow seas around land masses where animals and plants have died and have been compressed over millions of years. When extracted, it comes out in an unusable form known as **crude oil**. This needs to go through a refinery process to break it down into more useful components. These include petrol, diesel, oil for lubrication, paraffin oil and oil to make plastics.

Gas

Natural gas is found in conjunction with oil, as it is produced when the plants decay.

 DON'T FORGET

Make sure you know what types of energy are renewable and non-renewable, and that you can state several examples of each. You also need to know their advantages and disadvantages.

 THINGS TO DO AND THINK ABOUT

Copy and complete the table below, filling in as many examples as you can think of in each section.

Renewable energy	How is it harnessed?	Non-renewable energy	How is it used?

 ONLINE TEST

Head to www.brightredbooks.net to test yourself on energy.

ENERGY: THE GREENHOUSE EFFECT AND THE CONSERVATION OF ENERGY

THE GREENHOUSE EFFECT

A greenhouse is a house made of glass that is used to grow tomatoes, flowers and other plants in it. A greenhouse stays warm inside, no matter the temperature outside. Sunlight shines in and warms the plants and air inside, but the heat is trapped inside by the glass. The Earth's atmosphere does the same thing as the greenhouse. Gases that have been propelled into the atmosphere, such as carbon dioxide, do what a greenhouse does, and stop the heat escaping. These gases mainly exist through the burning of fossil fuels. This has triggered global warming, which has caused severe weather incidents such as flooding and droughts, a rising sea level and the melting of the polar ice caps.

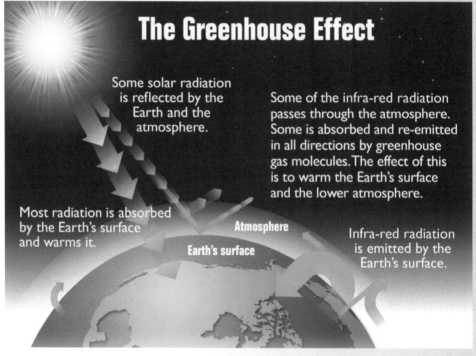

The Greenhouse Effect

Some solar radiation is reflected by the Earth and the atmosphere.

Some of the infra-red radiation passes through the atmosphere. Some is absorbed and re-emitted in all directions by greenhouse gas molecules. The effect of this is to warm the Earth's surface and the lower atmosphere.

Most radiation is absorbed by the Earth's surface and warms it.

Atmosphere

Earth's surface

Infra-red radiation is emitted by the Earth's surface.

ONLINE

Click onto the BBC Bitesize link at www.brightredbooks.net to get more information on global warming and the greenhouse effect.

ONLINE

Click onto the Practical Action link at www.brightredbooks.net to play Moja Island to reinforce your understanding of renewable energies.

VIDEO LINK

Check out the clip 'None like it hot' at www.brightredbooks.net for more on global warming.

ONLINE

Click onto the 'My Sustainable House' link at www.brightredbooks.net to play a game to strengthen your understanding of sustainability and energy-saving.

CONSERVATION OF RESOURCES

The law of conservation of energy is: '*Energy cannot be created or destroyed. It can only be changed from one form to another.*' But our current energy use is inefficient and dangerous for the planet. The best thing we can do to help save the Earth is to use and to waste less energy. This will put fewer greenhouse gases into the atmosphere, as well as other poisonous gases which can have dangerous consequences such as acid rain. Acid rain has a lower pH than normal, which can have harmful effects on plants, aquatic animals and infrastructure. In addition to this, it will help to save and prolong the fuels we have left.

Our modern industrialised society has been very wasteful of energy, but there are many ways in which we can improve. Engineers have been working on ways to combat climate change, and many solutions have been put in place to do this. Examples of this are the research into carbon capture and storage, the vast improvement in the efficiency of renewable-energy generators, and the increase in alternative powered transport.

contd

A carbon footprint is the measurement of the total greenhouse gas emissions that are caused directly and indirectly by a person or organisation. We are currently leaving behind a very large carbon footprint, and we need to change this for the sake of our planet. If we use more energy-efficient methods, we could save up to ¾ of all energy we use in our homes, offices, factories and modes of transport. Energy-saving is something that can be easily done, particularly at home. For example, we can reduce our carbon footprint by switching off unneeded electrical items overnight, or at least putting them into stand-by mode. If every Sky box in the UK was turned onto stand-by overnight, the amount of energy saved would be enough to power Birmingham for a whole year!

 ONLINE

Go to the 'WWF Footprint Calculator' link at www.brightredbooks.net to work out your carbon footprint and find out ways to reduce this.

 ONLINE TEST

Head to www.brightredbooks.net to test yourself on energy.

 DON'T FORGET

As a future engineer, your role is directly and indirectly about saving the planet. By knowing what your waste energy is, it allows you to make your engineering solutions more efficient by trying to reduce or remove these. The more efficient a design is, the more the client will save money, meaning they are more likely to use your solution.

 THINGS TO DO AND THINK ABOUT

Pick a type of renewable energy and complete these sentences.

- Renewable energy is good because ...
- But it has some problems, for example ...
- I think to meet Scotland's energy needs in the future we should ...
- I can reduce my carbon footprint by ...

 VIDEO LINK

Watch the video at www.brightredbooks.net to get a greater understanding of the need for sustainable energy.

ENERGY CALCULATIONS 1

To help reduce our energy, first we have to know how much energy we are using. There are several calculations we can use to work this out.

CALCULATING WORK DONE

When a force is used to move an object, 'work' is said to be done. An example of this is pushing a car from position A to position B.

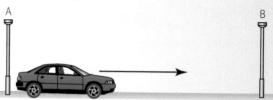

The amount of work you will do depends on two things: how difficult the car is to push (the size of the force) and how far it has to be pushed (the distance). The amount of work that has to be done can then be calculated using the following formula:

$$EW = F \times d$$

(work done = force applied × distance moved)

- **Force** is measured in **Newtons (N)**
- **Distance** is measured in **metres (m)**
- **Work** is measured in **joules (J)**.

EXAMPLE:

A lift raises a mass of 1500 kg to the next floor in a local shopping centre, which is 25 metres up. Calculate the minimum amount of work that must be done by the winch.

SOLUTION:

First, we have to work out the **weight**, as we are only given the **mass** of the elevator. Remember, weight is measured in N and mass is measured in kg.

Weight of lift = mg (mass × gravitational pull)

$$= 1500 \times 9{\cdot}8$$

$$= 14700\,N$$

... now we have the information to work out the **work done**.

E_w = force × distance

$$= 14700 \times 25$$

E_w = **367 500 J**

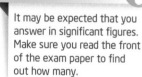

DON'T FORGET

The gravitational pull is always $9{\cdot}8\,ms^{-2}$. This is in your data booklet if you cannot remember it.

DON'T FORGET

It may be expected that you answer in significant figures. Make sure you read the front of the exam paper to find out how many.

⚙ ACTIVITY 1:

At a large distribution warehouse, products are stored on pallets and moved about by forklift trucks. If a forklift truck lifts a pallet of merchandise weighing 550 kg to a height of 3·5 m, calculate the minimum amount of work the truck must do during the lift.

CALCULATING POWER

Power is a measure of the rate of energy transfer. It gives an indication of how quickly the energy is changed from one form to another.

contd

Power is calculated using the following equation: **P = E/t**

power = energy transfer ÷ time

- **Power** is measured in **watts (W)**
- **Energy** transfer is measured in **joules (J)**
- **Time** is measured in **seconds (s).**

EXAMPLE:

If an LED light bulb uses 12 kJ of energy in 5 minutes, what is the power rating of the bulb?

SOLUTION:

$P = E/t$

$= 12\,kJ \div 5\,minutes$

$= 12\,000 \div 300\ (5 \times 60\ seconds)$

$= 40\,W$

CALCULATING ELECTRICAL ENERGY

Electrical energy is energy that is caused by moving electric charges. It is used to power a lot of things in your everyday life, from your TV to your microwave oven.

The formula for calculating electrical energy is: $E_e = VIt$

(electrical energy = voltage × current × time)

- **Voltage (V)** is measured in **volts (V)**
- **Current (I)** is measured in **amps (A)**
- **Time (t)** is measured in **seconds (s).**

DON'T FORGET

Always convert your time into seconds when using time!

EXAMPLE:

An electric hob in a school's home economics department has an operating voltage of 230 V with a current of 5 A. Calculate how much electrical energy will be used if the hob takes 7 minutes to boil a pot of soup.

SOLUTION:

$E_e = VIt$

$= 230 \times 5 \times (7 \times 60)$

$= 230 \times 5 \times 420$

$= 483\,000\,J$

$E_e = \mathbf{483\,kJ}$

 ACTIVITY 2:

A hot-air hand-dryer activates for 15 seconds once your hands are sensed. The dryer operates from a 220 V supply and draws a current of 12 A. Calculate the amount of electrical energy used when the dryer is operating.

THINGS TO DO AND THINK ABOUT

An Edinburgh tram can accelerate to its top speed of 15 ms⁻¹ in 10 seconds. During this acceleration, its electric motors draw a current of 1500 A at 750 V.

Calculate the energy going into the system.

Why would it be likely that the tram will actually produce this amount of output energy as it moves?

 ONLINE TEST

Head to www.brightredbooks.net to test yourself further on work done, power and electrical energy.

ENERGY CALCULATIONS 2

CALCULATING KINETIC ENERGY

All moving things have kinetic energy. Kinetic energy is the energy within an object when it moves. These can be very large things, such as planets, but also very small ones, like atoms. The heavier an object is, and the faster it moves, the more kinetic energy it possesses.

The formula for calculating kinetic energy is:

$E_K = \frac{1}{2} mv^2$

(Kinetic energy = ½ × mass × velocity²)

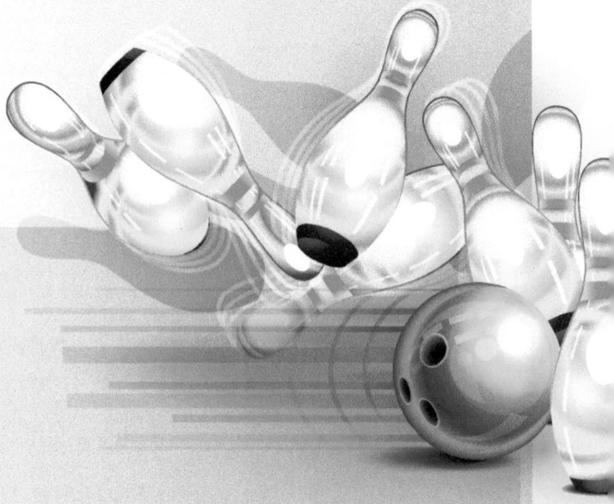

- **Mass (m)** of the object is measured in **kg**
- **Velocity (v)** is measured in **ms⁻¹**.

EXAMPLE:

If a 145 kg motorbike travels at 60 ms⁻¹, how much kinetic energy does it possess?

SOLUTION:

$E_K = \frac{1}{2}mv^2$

$= 0{\cdot}5 \times 145 \times 60^2 = 0{\cdot}5 \times 145 \times 3600$

$= 261\,000\,J$

$= \mathbf{261\,kJ}$

 **ACTIVITY 1:**

A man and a woman are both running at 3 ms⁻¹. The man has a mass of 76 kg and the woman has a mass of 63 kg. Work out the kinetic energy each person possesses.

 **ACTIVITY 2:**

A weight has been dropped and is falling at a speed of 4 ms⁻¹. If its kinetic energy is 34 kJ, what is its mass?

CALCULATING POTENTIAL ENERGY

Potential energy is energy that is stored within a stationary object that is capable of becoming active. For example, a ball on a high shelf has the potential to fall to the ground.

The formula used for calculating potential energy is:

$E_P = mgh$

(potential energy = mass × gravity × height)

- **Mass (m)** is measured in **kilograms (kg)**
- **Height (h)** is measured in **metres (m)**.

EXAMPLE:

A boy with a mass of 45 kg climbs 12 m up a tree. Calculate the potential energy he has when he reaches this height.

SOLUTION:

$E_P = mgh$

$= 45 \times 9{\cdot}8 \times 12 = 5292\,J$

$= \mathbf{5{\cdot}2\,kJ}$

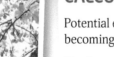

contd

⚙ ACTIVITY 3:

A rollercoaster has a highest point of 95 metres. When it reaches its summit, it pauses for riders to look over and see the drop. If its carriage is carrying a total mass of 1560 kg, what is its potential energy?

⚙ ACTIVITY 4:

A beam weighing 1500 kg is lifted by a crane. If its potential energy is 100 kN, how high has it been lifted?

CALCULATING HEAT ENERGY

Matter is made up of different molecules, and these molecules vibrate constantly. Heat energy is what we call the energy that comes from these vibrations. The hotter the substance, the more its molecules vibrate, and therefore the higher the heat energy.

The formula for calculating heat energy is:

$E_h = cm\Delta T$

(heat energy = specific heat capacity of material × mass × change in temperature)

- The **specific heat capacity (c)** of a substance is the amount of energy required to raise the temperature of 1 kg of the material by 1°C. This is measured in $Jkg^{-1}K^{-1}$.
- **Mass (m)** is measured in **kg**
- **Change in temperature (ΔT)** is measured in **Celsius (°C)** or **Kelvin (°K)**.

DON'T FORGET

The specific heat capacity of water is 4190 Jkg⁻¹K⁻¹. This will be in your data booklet if you cannot remember it.

EXAMPLE:

A large national building company has decided to install solar panels on the roofs of all its new houses.

How much energy is needed to heat 10 kg of water from a temperature of 10°C to a temperature of 80°C?

SOLUTION:

Change in temperature = 80 − 10

        = 70°C

$E_H = cm\Delta T$

   = 4190 × 10 × 70 = 2 933 000 J

   = **2·93 MJ**

⚙ ACTIVITY 5:

The water in a hot tub has a mass of 1500 kg. What is the heat energy required to raise the temperature from 15°C to 24°C?

⚙ ACTIVITY 6:

6·2 kJ of thermal energy is supplied to 19 kg of oil having a specific heat capacity of 2·7 Jkg⁻¹K⁻¹. If the initial temperature of the oil is 9°C, what will be its final temperature?

THINGS TO DO AND THINK ABOUT

A 0·9 kg ball is dropped from a height of 15 m.
(i) Calculate the potential energy of the ball from this height.
(ii) Calculate the maximum velocity of the ball.

Take your answers into class to discuss with your friends. Are you correct? How would your answers be affected if the ball weighed more?

ONLINE TEST

Head to www.brightredbooks.net to test yourself further on kinetic, potential and heat energy.

ENERGY TRANSFER AND LOSS

ENERGY TRANSFER

The rule of conservation states that energy cannot be created or destroyed. Instead, it can only be changed from one form of energy to another. How this energy is transformed, or converted, is very important to an engineer. Some forms of energy are directly interchangeable, but others need to go through several changes to arrive at the desired form. For example, consider Cruachan Hydro-electric Power Station.

Hydro-electricity

The water that is stored behind the dam contains **potential energy**. As it is released into the penstock, it changes into **kinetic energy**. As it flows through the turbine, its kinetic energy causes it to move, creating **electric energy** in the generator.

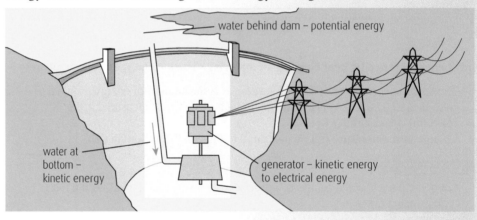

water behind dam – potential energy

water at bottom – kinetic energy

generator – kinetic energy to electrical energy

ONLINE

Visit
www.brightredbooks.net
to find more information
about the 'Hollow Mountain
Power Station' and how it
creates electricity.

Geothermal energy

Another example of harnessing energy that has to go through multiple conversions is **geothermal energy**. Geothermal energy is a renewable, clean and sustainable energy that utilises the heat stored in the Earth. This is becoming a more popular way of gathering energy, as it can be used on a domestic scale to heat your house, as well as on an industrial scale where it can be used to create electricity in power stations.

Within a geothermal power station, wells are drilled into the ground. Using **kinetic energy**, cold water is poured down into them. This water is changed through the earth's **heat energy** into high-pressure steam, which is brought back to the surface through another well. As the steam rises to the surface, it turns the blades of a turbine, producing **kinetic energy**, which in turn powers a generator to create **electrical energy**.

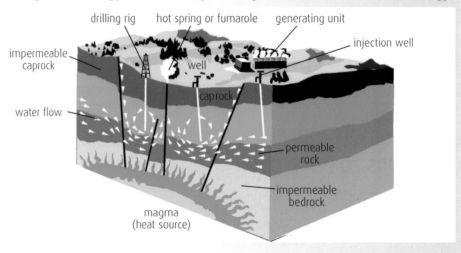

drilling rig   hot spring or fumarole   generating unit

injection well

impermeable caprock

well

caprock

water flow

permeable rock

impermeable bedrock

magma (heat source)

ENERGY LOSSES

We all know that energy cannot be destroyed, and that the energy output from a system must equal its input; but not all the energy will be transferred in the way we want. When an energy conversion takes place, there is always a change that we do not need or want. This is known as waste energy. This usually comes in the form of heat or sound from any moving parts of a mechanism.

If we look at a simple energy conversion, we can expand it to show the waste energy, or energy losses. For example, our earlier example of a school bus:

The bus will change chemical energy (petrol or diesel) into kinetic energy (movement) – but, in doing so, heat and sound will be created by the friction and movement in the mechanisms throughout the system.

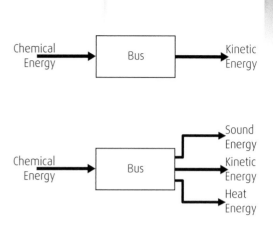

CALCULATING ENERGY TRANSFERS

During any energy transformation, the total energy contained within must remain constant at all times. Knowing the total amount of energy at the start of any energy transformation, we should know how much energy it is using throughout the whole process.

EXAMPLE:

For example, if a body of mass 50 kg falls freely from a height of 30 metres, we can find out the total energy throughout the mass at all times.

SOLUTION:

First, we calculate the potential energy it holds at the start.

E_p = mgh

$\quad$ = 50 × 9·8 × 30

$\quad$ = 14 700 J

$\quad$ **= 14·7 kJ**

This means we now know the total amount of energy that this object can contain. From this, we can transfer the energy into any of our other equations to work out different factors. For example, we could take this energy and transfer it into the kinetic-energy equation to work out the final velocity at impact.

$E_k \quad = \frac{1}{2} mv^2$

$14\,700 = 0·5 × 50 × v^2$

$14\,700 = 25 × v^2$

$v^2 \quad = \frac{14\,700}{25}$

$v^2 \quad = 588$

$v \quad = \sqrt{588}$

$v \quad = 24·25\,ms^{-1}$

THINGS TO DO AND THINK ABOUT

A travel kettle takes 4 minutes to boil water. If the original temperature of the water is 15°C and the kettle contains 1·2 litres of water, calculate how much electrical current is used when it is powered from a European 220 V mains supply.

Research three different types of renewable energy and how they are harnessed. Are there any types that could easily be used domestically? Bring your findings into school and discuss with your class.

ONLINE TEST

Test yourself on energy transfer and loss at www.brightredbooks.net

ENERGY EFFICIENCY AND AUDITS

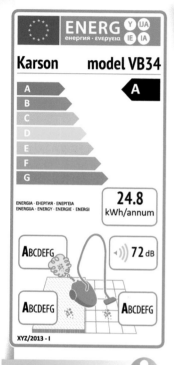

Karson model VB34

ENERGIA · ЕНЕРГИЯ · ЕΝΕΡΓΕΙΑ
ENERGIJA · ENERGY · ENERGIE · ENERGI

24.8 kWh/annum

72 dB

XYZ/2013 - I

DON'T FORGET

As η is the ratio of output to input energy, it can never be greater than 1. When multiplying by 100, it will then give us the efficiency percentage.

ENERGY EFFICIENCY

Energy efficiency is about using less energy to provide the same amount of power. Energy efficiency is something you have probably heard about, as it is something we do in our own houses, as well as it being something that industry has been considering for a long time. By making products and processes more energy-efficient, not only will it help reduce waste energy, therefore saving money, but also it will help reduce the greenhouse gases that are emitted and will therefore lower our carbon footprint. Domestically, it is done by making your household more efficient in its energy usage with the aim of reducing your household bills.

Becoming energy-efficient is something you will be doing already and you may not even realise it. Do you have energy-saving, LED or halogen light bulbs in your house? Have your parents bought any white goods, a house or a car recently? These are all rated on their efficiency to make the buyer more aware of their energy usage.

It is possible to look at how well an energy system is operating by calculating its efficiency. As engineers, we need to know the efficiency of an energy transformation to make sure our engineering solutions are as efficient as possible. The more efficient they are, the more effective they will be. By comparing the input energy and the useful output energy, we can make our designs better.

This can be calculated using the following equation:

$\eta = E_{out} / E_{in} \times 100\%$

(efficiency = useful energy output / total energy in)

EXAMPLE:

A coffee machine heats one cupful of water (0·15 kg) from 20°C to 85°C in 15 seconds. The heating element operates from a 230 V, 22 A supply.

a) Calculate the electrical energy going into the coffee machine.

SOLUTION:

$E_e = VIt$

$= 230 \times 22 \times 15$

$= 75\ 900\ J$

$= 75·9\ MJ$

b) Calculate the heat energy transferred to the water.

SOLUTION:

$E_H = cm\Delta T$

$= 4190 \times 0·15 \times (85 - 20)$

$= 40\ 852·5\ J$

$= 40·9\ MJ$

c) Calculate the efficiency of the machine.

SOLUTION:

$\eta = E_{out} / E_{in}$

$= 40·9 / 75·9$

$= 0·539 \times 100\%$

$= 53·9\%$

ENERGY AUDITS

As we know, it is not possible to 'destroy' energy – it must go somewhere, and this is known as waste energy. The problem with any engineering solution is that not all energy going into the system will come out as a useful form of energy. For example, let's consider a light bulb. This turns electric energy into light energy, but in doing so it also creates heat – heat is the waste energy in this system.

Since we know that the total energy in any closed system must be constant, we can carry out meaningful calculations if we remember to take all types of input and output energy into account.

In order to ensure that we have taken all energies into account, it is useful to carry out an energy audit. An energy audit is a list of all energies coming INTO and going OUT OF a system. The total for the energies IN **must** be the same as the total for the energies OUT.

To do this, we work out the efficiency, and then consider as waste energy any other energies potentially created.

> **EXAMPLE:**
>
> If we consider the coffee machine that had its efficiency calculated, we know it has an input of electrical energy worth 75·9 MJ, and an output of heat energy worth 40·9 MJ. This means that it is only 53·9% efficient, and 46·1% of all energy is created is waste energy. It could be assumed that a lot of the waste energy takes the form of sound. This information can now be drawn into an energy audit diagram.
>
>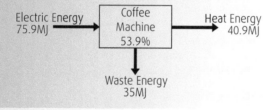

 ONLINE TEST

Head to
www.brightredbooks.net
to test yourself further on
efficiency and energy audits.

 THINGS TO DO AND THINK ABOUT

When a maintenance engineer tests a 12 V, 3 A car-tyre pump, they find that it produces 220 J of energy over a period of 10 seconds. Work out the input energy for the system and the efficiency of the system, and then draw your results in the form of an energy audit diagram.

Take your working-out into class to discuss, and to discover if you have got it correct.

ANALOGUE ELECTRONICS: ELECTRICAL AND ELECTRONIC ENGINEERING

WHAT IS ELECTRICAL AND ELECTRONIC ENGINEERING?

As we know, engineering is vital to everyday life – it shapes the world we live in and helps to create its future. Electrical and electronic engineering in particular are exciting branches that are at the cutting edge of engineering. They use scientific knowledge of the behaviour and effects of electronics to design and test systems, devices and equipment. In recent years, we have seen the introduction of smartphones, tablet computers, hybrid cars and 3-D printers. All of these technologies are now commonplace in today's society. These, plus many more things, have all been created by electronics engineers.

Although it could be argued that they are very similar and will work together a lot due to their overlapping interests, an **electrical** engineer focuses on power generation, transmission, utilisation and measurement. **Electronics** engineers deal with automatic control of systems and the implementation of it. They will work with analogue and digital circuits as well as computer programming. Basically, electrical engineers deal with the electricity, electronics engineers deal with the manipulation of it.

VIDEO LINK

Check out the clip on electronic engineering at www.brightredbooks.net

Electronics and electrical components are the basis for all modern technology, and it has advanced significantly in your lifetime. If you think of your first mobile phone and compare it to your current one, your old phone wouldn't have been able to do half the stuff your current one could!

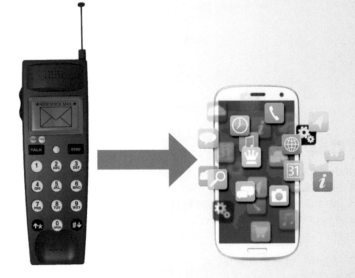

DON'T FORGET

An electrical or an electronics engineer is not the same thing as an electrician. An engineer will design the electrical system, whereas an electrician will install it. Do not get these mixed up when answering questions in an exam!

Within the National 5 Engineering Science course, we will study three different branches of electrical and electronic engineering – analogue electronics, digital electronics and programmable control.

ELECTRONIC CIRCUITS

An electronic circuit is made up of electrical components such as batteries, bulbs and switches. It is always within a closed-loop circuit, allowing current to flow connecting to the positive and negative parts of a voltage supply.

Electric current is the name given to the flow of negatively charged particles called electrons.

contd

Current (I) is measured in **amperes (A)**. The current is the rate of flow of the electrons through the circuit.

Voltage (V) is used to drive the electrons through the components within the circuit. Voltage is measured in **volts (V)**.

Resistance (R) is the measure of how much a material reduces the amount of current flowing through it. Resistance is measured in **ohms (Ω)**.

DON'T FORGET

All symbols for units and quantities can be found in your data booklet, so make sure you use this if unsure.

If we think of an electrical circuit as being like a river, the voltage is how deep the river is, the current is how fast the river is flowing, and the resistor is a dam that is only allowing a certain amount of water through. Increasing the resistance will decrease the current that may be driven through the circuit by the voltage.

Ohm's Law

Ohm's Law is the relationship in a circuit between the three electrical variables: voltage, current and resistance.

$V = IR$

voltage = current × resistance

$V = I \times R$ $I = \dfrac{V}{R}$ $R = \dfrac{V}{I}$

VIDEO LINK

Watch the clips at www.brightredbooks.net to help develop your understanding of Ohm's Law.

Power in a circuit

Electrical **power (P)** is the rate at which energy is used within a circuit. This is measured in **watts (W)**.

The power in an electric circuit depends both on the amount of current (I) flowing through it and on the amount of voltage (V) that is applied. To calculate the power in a circuit, you can use the following rule.

$P = IV$

$P = I \times V$ $I = \dfrac{P}{V}$ $V = \dfrac{P}{I}$

VIDEO LINK

Check out the videos at www.brightredbooks.net for more on using a scientific calculator for decimal prefixes.

 THINGS TO DO AND THINK ABOUT

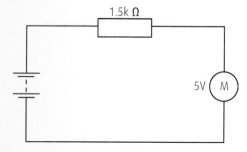

ONLINE TEST

Head to www.brightredbooks.net to test yourself on analogue electronics.

The current flowing through the resistor is 0.9 A.

a Calculate the voltage within this component.

b Calcuate the power in the motor.

ANALOGUE ELECTRONICS: ELECTRONIC SYMBOLS AND RESISTOR COLOUR CODES

ELECTRONIC SYMBOLS

Power supplies

Name	Symbol	Function	Purpose
Battery		Power supply	To power a circuit
DC supply voltage		Power supply	To power a circuit
Resistor		Passive component	Reduces the amount of current flowing through a circuit. Resistance is measured in **ohms (Ω)**.
Variable resistor		Input component	Changes the amount of resistance in a certain part of a circuit
Light-dependent resistor		Input component	Changes the resistance depending on the light level
Thermistor		Input component	Changes the resistance depending on the temperature
Diode		Semiconductor device	Only allows electricity to flow in 1 direction
Transistor		Semiconductor device	Acts as a digital switch. When 0·7 V passes through, it saturates and switches on.
Relay		Electromagnetic switch	Used to connect a circuit to another circuit with a far larger power supply
Light-emitting diode		Output component	Will glow to create light energy when electricity flows through it
Motor		Output component	Will turn to create kinetic energy when electricity flows through it
Lamp		Output component	Will glow to create light energy when electricity flows through it
Ammeter		Measuring device	Used to measure the current going through the circuit
Voltmeter		Measuring device	Used to measure the voltage going through a certain part of the circuit

RESISTOR COLOUR CODES

As we know, resistors are used for regulating current. They 'resist' the current flow, and the amount of resistance they give is measured in ohms (Ω). Resistors are found in almost every electronic circuit to ensure components do not get too much current through them, as this would cause them to blow.

All resistors are marked with coloured bands to show how much resistance they have. Each band of colour helps to identify the value (in ohms) and its tolerance (in percent). The tolerance is there to allow for a certain amount of leeway, allowing it to deviate positively or negatively from the set value.

The colour-code chart for resistors is shown below.

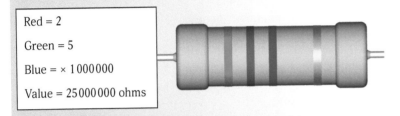

Red = 2

Green = 5

Blue = × 1 000 000

Value = 25 000 000 ohms

Colour	1st band	2nd band	3rd band (multiplier)	4th band (tolerance)
Black	0	0	× 1	
Brown	1	1	× 10	± 1%
Red	2	2	× 100	± 2%
Orange	3	3	× 1000	
Yellow	4	4	× 10 000	
Green	5	5	× 100 000	
Blue	6	6	× 1 000 000	
Violet	7	7		
Grey	8	8		
White	9	9		
Silver			× 0·01	± 10%
Gold			× 0·1	± 5%

THINGS TO DO AND THINK ABOUT

Use the colour-code table to work out the colours for the following resistors:

- 2K9 (2900)
- 1M4 (1 400 000)
- 67
- 450
- 12K (12 000)
- 380

Get your fellow classmates to check your answers and discuss if you were right or wrong.

DON'T FORGET

You will not be expected to remember these, but it is worthwhile making sure you understand how the banding works. It will help you a lot with the practical aspects of this unit.

VIDEO LINK

Watch the video at www.brightredbooks.net to get a greater understanding of how resistor bands work.

ONLINE TEST

Head to www.brightredbooks.net to test yourself on electronic symbols.

ANALOGUE ELECTRONICS: TYPES OF CIRCUITS

There are two forms of circuit – **series** and **parallel**. A series circuit works with all components joined up together in one continuous loop for the electrons to flow around. A parallel circuit is broken up into different branches, with electrons having a 'choice' where to go.

SERIES CIRCUITS

A series circuit is the simplest form of circuit, but it has a major disadvantage. Within this type of circuit, if one component breaks, the loop will be broken with nowhere for the electrons to flow. This means all other components in the circuit will stop working.

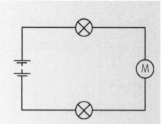

Voltage in a series circuit

Within a series circuit, the voltage is divided up between all components – this is known as Kirchhoff's Second Law. This states that:

$$V_{cc} = V_1 + V_2 + V_3 + ...$$

This can be proven by adding voltmeters to the circuit.

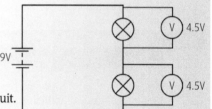

Current in a series circuit

Unlike voltage, which is divided in a series circuit between the components, the current flows through all of the components equally. It does not matter where you put the ammeter in a series circuit, as it will give you the same reading everywhere.

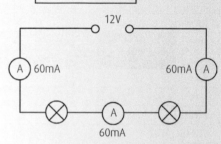

Resistance in a series circuit

As resistors come in standard sizes, they are often connected in series to obtain a specific size that is required. To work out the resistance in a series circuit, add the resistance of each resistor or component together, using the following equation.

$$R_{total} = R_1 + R_2 + R_3 ...$$

> **EXAMPLE:**
> $$R_{total} = R_1 + R_2 + R_3$$
> $$= 1K + 37 + 250$$
> $$= 1000 + 37 + 250$$
> $$R_{total} = 1287\,\Omega$$

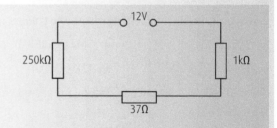

PARALLEL CIRCUITS

A parallel circuit differs from a series one in that the circuit is broken up into different branches. This allows the current to flow through each branch separately. This means that if one component fails, only the branch where this component is placed will stop working, as current can still flow through the rest of the circuit.

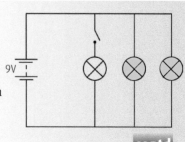

contd

Voltage in a parallel circuit

Within a parallel circuit, each branch receives the supply voltage. This means that you are able to run a number of devices from one supply, and they will all receive the same input voltage.

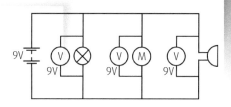

Current in a parallel circuit

As we know, each branch of a parallel circuit receives the total supply voltage – but this is not the same for current. When the current reaches a junction, it splits up, with some current going along one branch, and the rest going along the other branch.

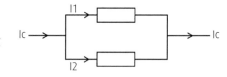

$Ic = I_1 + I_2 + ...$

Resistance in a parallel circuit

As we know, resistors come in standard sizes, but one way of obtaining a specific size that is otherwise unavailable is by connecting them in parallel.

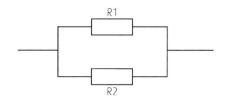

To calculate the total resistance in a parallel circuit, you use one of the following two ways:

$$R_t = \frac{R_1 \times R_2}{R_1 + R_2} \quad \text{or} \quad \frac{1}{R_t} = \frac{1}{R_1} + \frac{1}{R_2}$$

The first equation can only be used for two resistors in parallel. The second one is to be used for any amount of resistors connected in parallel.

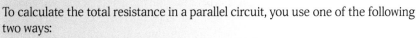

EXAMPLE:

Equation type 1

$$R_t = \frac{R_1 \times R_2}{R_1 + R_2} = \frac{1000 \times 250}{1000 + 250} = \frac{250\,000}{1250} = \mathbf{200\,\Omega}$$

Equation type 2

$$\frac{1}{R_t} = \frac{1}{R_1} + \frac{1}{R_2} = \frac{1}{1000} + \frac{1}{250} = \frac{1}{1000} + \frac{4}{1000} = \frac{5}{1000}$$

$$\frac{1}{R_t} = \frac{5}{1000} = \frac{R_1}{1} = \frac{1000}{5} = \mathbf{200\,\Omega}$$

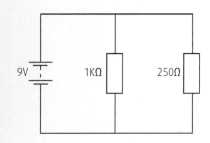

ONLINE TEST

Head to www.brightredbooks.net to test yourself on series and parallel circuits.

THINGS TO DO AND THINK ABOUT

Work out the resistance in the following circuits.

a

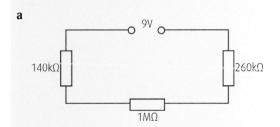

b

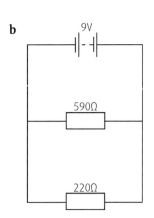

c

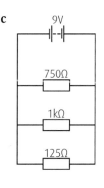

Get your fellow classmates to check your answers and discuss as a group if you were right or wrong.

ANALOGUE ELECTRONICS: INPUT TRANSDUCERS

Input transducers are input devices that are used to convert a change of physical conditions into a change in resistance and/or voltage. Examples of this are thermistors (in which resistance changes by temperature), LDRs (in which resistance changes by light) and variable resistors (in which the resistance can be changed by turning a dial).

To work out the resistance for these, we can use a graph.

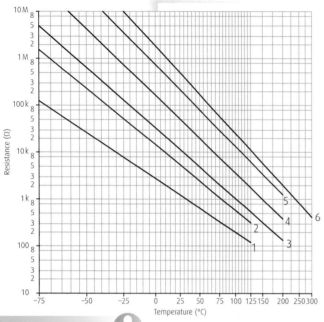

THERMISTOR GRAPH

The question will tell you which type of thermistor it is, for example a type 2 thermistor. There are six diagonal lines on the graph, and each one refers to a different type of resistor. The question will then ask you to work out the resistance by giving you the temperature, or vice versa.

If a question in an exam asks you to work out the resistance/input conditions for one of these components, the graph will be shown in the question.

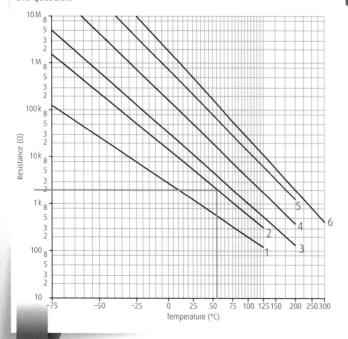

EXAMPLE:

What is the resistance within a type 2 resistor if the temperature is 55°C?

SOLUTION:

To answer this, find 55°C on the temperature axis. The number it is asking for may not necessarily be shown in the graph, but you can use problem-solving skills to work out where this is. Lines for 50°C and 75°C are marked, with 4 lines between them. You can then deduce that this means 5°C per line, meaning you could find 55°C.

Then follow the line up until it hits the diagonal line associated with this type of thermistor, and take this line across to the y-axis to find the resistance.

This is now hitting just below the 2 on the resistance axis. This 2 is in the 1 k section, so that means the resistance is just below 2000 Ω.

Each section of this axis is broken up into decimal values. If you look at the axis, it starts at 10, and then there are the numbers 2, 3, 5 and 8. These actually mean 20, 30, 50 and 80, as they are within the 10s. The next area is 100, with the same numbers: 2, 3, 5 and 8. In this case, they represent the numbers 200, 300, 500 and 800. This continues all the way up.

LDR GRAPH

The same technique used for thermistor graphs can be used for LDR graphs.

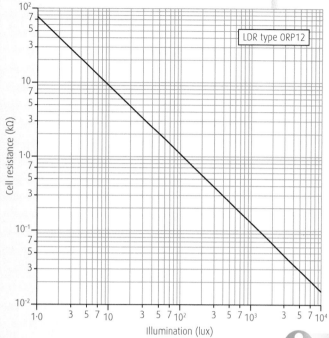

By knowing the lux (light level), we can then work out the resistance, and vice versa.

EXAMPLE:

What is the light level if an LDR has a resistance of 2 kΩ?

SOLUTION:

We firstly have to find the information it has given us on the graph. The axis with resistance values states that it is measured in kΩ, so we have to find 2. Once again, this is not always obvious, so we use our problem-solving skills to find it.

Once we have done this, we take this along to our diagonal line (in this case there is only one, as it is for a specific type of resistor – ORP 12), then follow it down to our bottom axis.

This falls directly onto the 5, but it is within the 10s section. This means our lux level is 50.

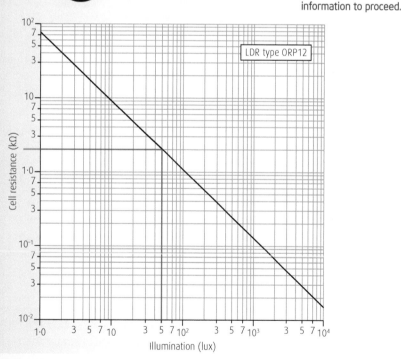

THINGS TO DO AND THINK ABOUT

Using the graph above, work out the following:
a   What is the resistance if the lighting is 300 lux?
b   What is the temperature on a type 2 thermistor if it is 20 000 Ω?
c   What is the resistance on a type 2 thermistor if it is 100°C?
d   What is the light level if the resistance in an LDR is 30 kΩ?

ANALOGUE ELECTRONICS: VOLTAGE-DIVIDERS

VOLTAGE-DIVIDER CIRCUITS

As we know, an input transducer changes a physical parameter into an electrical signal as the physical conditions change. These are commonly a variation of a resistor – for example, an LDR or a thermistor, which change light and temperature respectively into resistance. To do anything meaningful, we then have to convert this resistance change into a voltage change. This way, the signal can be processed. This is usually done using a voltage-divider circuit. A typical voltage-divider circuit is shown here.

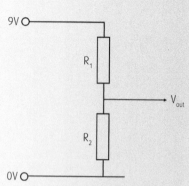

A voltage-divider circuit is basically two resistors connected in series, with the output coming from in between the resistors. If the value of resistor 1 is changed, the voltage across it will therefore have to change (think V = IR). This means the voltage across resistor 2 must also change. The input voltage will now be split between the two resistors – this is why it is known as a voltage-divider.

The construction of R_1 and R_2 can change to create different outputs. For example, alongside is shown to be a light sensor.

As the light level goes up, the resistance across the LDR drops, meaning the output will switch on when the light level rises. The variable resistor is there to control the sensitivity of the circuit. Basically, by changing the variable resistor, it will alter the amount of light needed to make the output go on.

If the two resistors are swapped around, the purpose of the circuit changes.

In a circuit such as this, it will become a dark-sensing circuit, and the output will go on depending on how dark it is. In a real-world scenario, this might be used for lighting in a residential care home, where the outside lights will go on automatically when it is dark and will therefore not be reliant on the occupants to switch them on.

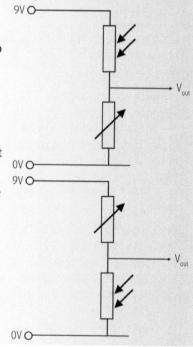

VOLTAGE-DIVIDER CALCULATIONS

To work out the output in this type of circuit, we use the following equation: $V_{out} = \frac{R_2}{R_1 + R_2} \times V_{cc}$

EXAMPLE:

$$V_{out} = \frac{R_2}{R_1 + R_2} \times V_{cc}$$

$$V_{out} = \frac{1000}{250 + 1000} \times 12$$

$$= \frac{1000}{1250} \times 12$$

$$= 0 \cdot 8 \times 12$$

$$\mathbf{V_{out} = 9 \cdot 6\,V}$$

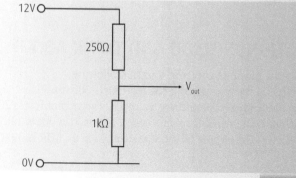

contd

⚙ ACTIVITY 1:

Calculate V_{out} for the following circuits.

(a)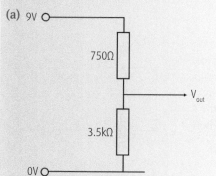

(b)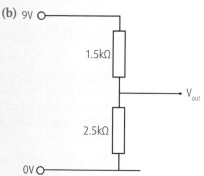

(c)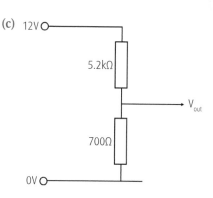

VOLTAGE-DIVIDER CALCULATION FOR COMPONENTS

If we have unknown quantities stopping us from working this out, we can also use this equation: $\frac{V_1}{V_2} = \frac{R_1}{R_2}$

EXAMPLE:

We know R_1 and V_2, but we can also easily work out V_1. If 3 V is running through resistor 2, and the supply voltage is 10 V, then that means the voltage through resistor 1 must be 7 V.

$$\frac{V_1}{V_2} = \frac{R_1}{R_2}$$
$$\frac{7}{3} = \frac{700}{R_2}$$
$$2{\cdot}3 = \frac{700}{R_2}$$
$$R_2 = \frac{700}{2{\cdot}3} = \mathbf{304\,\Omega}$$

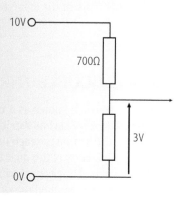

DON'T FORGET

Give your answers in significant figures, as this is what will be expected in the exam! If the question does not state the significant figures to use, then use the same amount used within the values in the question.

VIDEO LINK

Watch the clip at www.brightredbooks.net to get more information on how a voltage-divider circuit works.

⚙ ACTIVITY 2:

Calculate the missing component value for the following circuits.

(a)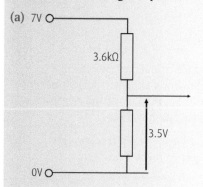

(b)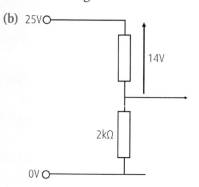

(c)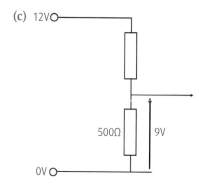

❗ THINGS TO DO AND THINK ABOUT

Think of some real-world applications for a voltage-divider circuit. Sketch out the circuit for these, using the proper electrical symbols.

ONLINE TEST

Head to www.brightredbooks.net to test yourself on voltage-dividers.

ANALOGUE ELECTRONICS: SWITCHES

SINGLE POLE, SINGLE THROW

As we know, a switch is used to deliberately break (or make) a circuit. The circuit shown here is of a simple series circuit that uses a switch, or more specifically a 'single pole, single throw' (SPST) switch. When the switch is down (on), the bulbs will come on, and when the switch is up (off), the bulbs will go off.

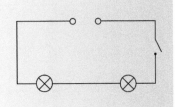

SINGLE POLE, DOUBLE THROW

The circuit here shows a type of switch known as a 'single pole, double throw' (SPDT) switch. This has one input (pole) with 2 output paths (throws). This will allow you to control two different possible paths, but both will not work at the same time.

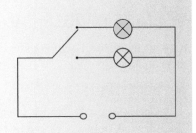

DOUBLE POLE, DOUBLE THROW

The circuit here uses a switch known as a 'double pole, double throw' (DPDT). This has two inputs (poles) and two outputs (throws) for each input. These are mainly used to control the direction of a motor. When it is switched one way, the motor will rotate clockwise. When switched the other way, it will rotate anti-clockwise. Both switches will always move at the same time.

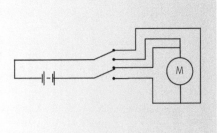

DIGITAL SWITCHES

Transistors

A transistor is a semiconductor device that uses small electrical currents to control much larger currents. It is a component that is a fundamental and ever-present part in all modern electronics. It could be argued that its invention revolutionised electronics and made it possible for us to have the electronic devices we have today.

A transistor has three connections underneath it – the base, the emitter and the collector. It is important that, when connecting this, you connect it properly.

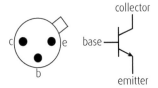

View from underneath

DON'T FORGET

A transistor acts as a switch and saturates at 0·7 V. We need a resistor before the base to protect it from too much current. You will need to know this for exams if asked to describe how this part of a circuit works.

In the context of this course, we use a transistor as a digital switch. The current flowing through the collector and emitter is controlled by the base. As soon as the base reaches 0·7 V, it 'saturates', amplifying the current and allowing it to flow into the collector and switch on the bulb.

contd

EXAMPLE:

To understand how a transistor works, think of it like a valve controlling the water flow in a pipe. When the input to the transistor gets 0·7V to the base (its input leg) it will saturate (switch on) and open the valve allowing the current to flow from the collector to the emitter. This will complete the circuit and allow the output device to switch on.

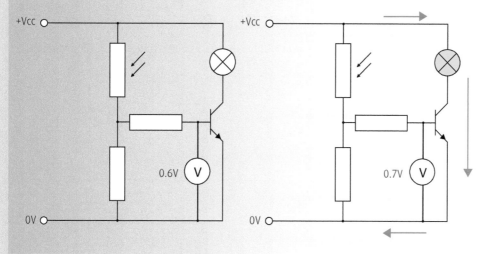

Relays

Although relays are often considered to be output devices, they are actually a type of electromagnetic switch. They are used mainly to join two circuits together, allowing a low-powered circuit to control another one that has a far higher power supply.

A relay has a coil that is energised and de-energised as the relay switches on and off. As the current flows through the coil, it creates a magnetic field. This causes it to attract the switch lever, pulling it and causing it to switch positions, joining the circuit.

During this process, the coil can generate a large amount of voltage called 'back EMF'. This can flow through the circuit the wrong way, causing considerable damage to components, especially transistors. To stop this from happening, we can place a diode over the power connections of the relay. Placing a diode in the circuit can block this reverse voltage and provide a path for it to escape.

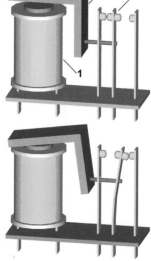

Relay principle: **1.** Coil, **2.** Armature, **3.** Moving contact

Relays are mainly used with SPST switches and DPDT switches.

As an SPST relay, it can be used to control any simple system, such as a lamp, or a motor that needs a higher power supply.

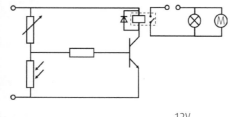

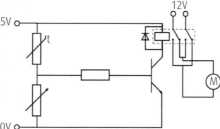

As a DPDT relay, it is ideal for controlling a motor. By using it in this configuration, it can make the motor turn one way; and when the relay is switched on, the direction of the motor will change.

THINGS TO DO AND THINK ABOUT

Breaking the two relay-based circuits above into input, process and output, describe how each of them works.

DON'T FORGET

Make sure you remember about back EMF in case you are asked in your exam to describe circuits that contain a relay.

VIDEO LINK

Check out the clip at www.brightredbooks.net to get more information on how a relay works.

ONLINE TEST

Head to www.brightredbooks.net to test yourself on switches.

ANALOGUE ELECTRONICS: COMPLETE CIRCUITS

DESCRIBING HOW A CIRCUIT WORKS

Within the exam, it is very likely you will be shown an electronic circuit diagram and asked to describe how it works.

> **EXAMPLE:**
>
> An electronics engineer has designed a circuit for electronic window blinds. Describe how the circuit works.
>
> Do not let this complex circuit confuse you. Just break it up into different sub-systems – input, process and output. If you find it easier, you can even title mini-sections to do this – the markers will not judge your format, they will only look at the content.

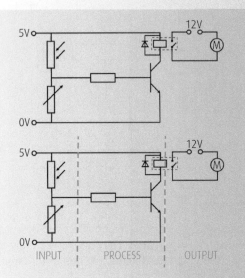

Input

This is potentially worth up to 2 marks: 1 for describing how the voltage-divider works, and another for saying why a variable resistor is included. Writing something like this would get you full marks:

- As the light level increases, the resistance of the LDR decreases. This also decreases the voltage dropped across it. This also means that the voltage across the variable resistor (V_{in}) increases.
- The sensitivity of the circuit can be adjusted by the variable resistor.

Process

This would potentially be work up to 3 marks: 1 for describing how the transistor works, 1 for stating the need to put a resistor in front of it, and another for describing how the relay works and the need for a diode. Writing something like this would get full marks:

- The transistor will saturate (switch on) when the light level reaches a set level and the voltage reaches 0·7 V.
- A resistor is placed before the transistor to protect it and stop too much current from flowing to the base.
- The relay switches on when the transistor saturates. A diode is present to protect the transistor and the circuit from back EMF.

Output

It is more than likely that this section would only be worth 1 mark for describing what happens when the relay is powered. Writing something like this will achieve this mark:

- When the relay is powered, it will close the switch, completing the larger voltage circuit. This will then power the motor, causing it to move.

Although a question like this may not be worth 6 marks in the exam as suggested, you should always try to describe the circuit fully in order to ensure you achieve full marks.

contd

EXAMPLE:

An electronics engineer has designed the circuit below to control an electronic toy car for small children.

Using the appropriate terminology, describe how the circuit works.

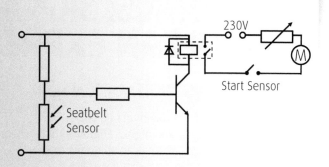

Input

- The voltage-divider circuit will act as a dark sensor. When the child is seated and the seatbelt is on, the LDR will sense darkness.
- The resistance and voltage across the LDR will increase.

Process

- When the transistor receives 0·7 V to its base, it will saturate (switch on).
- A resistor is placed before the transistor to protect it and to stop too much current from flowing to the base.
- The relay switches on when the transistor saturates. A diode is present to protect the transistor and the circuit from back EMF.

Output

- When the relay is powered, it will close the switch, switching on the 230 V circuit.
- When the driver presses the start switch, the motor will start to turn.
- The variable resistor can be turned to change the speed of the motor.

ACTIVITY:

(a) A circuit that is to be used in a house's central heating is shown here.

Using the appropriate terminology, describe how this circuit works.

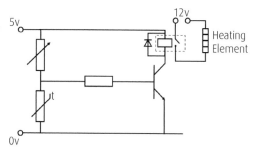

(b) The circuit here has been designed to be used in a runner's headlamp.

Using the appropriate terminology, describe how the circuit would work.

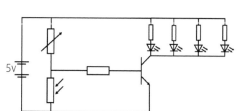

THINGS TO DO AND THINK ABOUT

Try to draw your own circuit, and get your classmates to work it out. Ask them to do the same for you, and work on circuit descriptions together.

ONLINE TEST

Test yourself on this topic at www.brightredbooks.net

DIGITAL ELECTRONICS: SIGNALS, LOGIC GATES AND BOOLEAN EXPRESSIONS

ANALOGUE AND DIGITAL SIGNALS

Electronic signals can be broken up into two distinct types – analogue and digital. Digital signals **must** be either on or off, whereas an analogue one can be in between. Think of a radio: if it starts to lose signal, the sound will become distorted and will start to hiss. This is because it works with analogue signals and it will still produce sound even though the signal is weak (it is between on and off). On the other hand, with a DAB digital radio, if you start to lose the signal the sound will switch off.

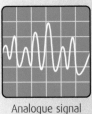

Analogue signal

Digital signal

DON'T FORGET

It is worth memorising these and making sure you understand how they work – this is the basis for all digital electronics.

LOGIC GATES

Logic gates are the basic building blocks of a digital circuit and are used to process a combination of different inputs.

There are three different types of logic gates you will need to know for the National 5 course: AND, OR and NOT. Below is the symbol for each gate, as well as its truth table. A truth table is a table made up of binary 1s and 0s: 1 means the signal is on, and 0 means it is off. Basically, it tells the truth of what is happening in the circuit.

NOT gate

Within a NOT gate, the output is **NOT** the same as the input.

Input ——————▷○—— Output

A	Z
0	1
1	0

AND gate

Within an AND gate, both input A **AND** input B need to be on for the output to go on.

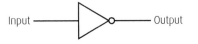

A ——————
B —————— Output

A	B	Z
0	0	0
0	1	0
1	0	0
1	1	1

OR gate

Within an OR gate, input A **OR** input B needs to be on for the output to go on.

A ——————
B —————— Output

A	B	Z
0	0	0
0	1	1
1	0	1
1	1	1

DON'T FORGET

It is worth memorising these Boolean expressions, as you should expect Boolean equations to appear in an exam question.

BOOLEAN EXPRESSIONS

Each logic gate has a corresponding mathematical equation. This is known as the Boolean expression.

NOT gate $Z = \bar{A}$ AND gate $Z = A.B$ OR gate $Z = A + B$

COMBINATIONAL LOGIC – BOOLEAN EXPRESSIONS

DON'T FORGET

Don't fall into the trap of using a '+' for AND, as this will give you the wrong answer. AND is '.' and OR is '+'.

In electronic systems and products, there will usually be a combination of several different logic gates, and it will rarely be as simple as what you have just learned. This is known as 'combinational logic'. If you know what the circuit looks like, you can then work out the overall Boolean expression for the circuit.

EXAMPLE:

Work out the Boolean expression for the logic circuit shown right.

To work this out, you have to break it down into different sections, working from left to right, and then work out the equation as it goes through each logic gate.

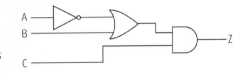

Step 1

You can find out that the line for A soon turns into Ā.

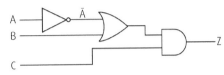

Step 2:

The next logic gate we come across is an OR gate. The paths into this gate are now Ā and B.

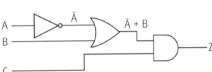

Step 3:

The logic gate after this is an AND gate. The paths we now have to enter are known as (Ā + B) and C.

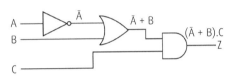

This means our Boolean expression for this circuit will be **Z = (Ā + B).C**

 ACTIVITY:

Work out the Boolean expressions for the circuits below.

(a)

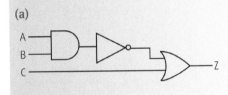

(b)

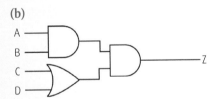

(c)

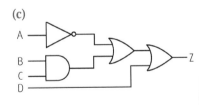

 THINGS TO DO AND THINK ABOUT

Use Yenka to build up several different combinational logic circuits. Use a lamp as the output and switches as the inputs. Try to work out how to switch them on, then test to see if you are correct. Write an evaluation of the circuits, stating **exactly** what you did to test each one. If you had to modify the circuit or your answers in any way, explain why.

 VIDEO LINK

The clip at www.brightredbooks.net provides further explanation of how to work out Boolean expressions.

 ONLINE TEST

Head to www.brightredbooks.net to test yourself on Boolean expressions.

DIGITAL ELECTRONICS: TRUTH TABLES 1

COMBINATIONAL LOGIC: TRUTH TABLES

For each combinational logic diagram, a logic table can also be worked out.

EXAMPLE:

First, we have to figure out how many rows and columns our truth table will need. To work out the columns, we have to name every single input and output to each gate.

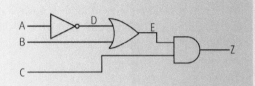

1 So, our logic table will have the columns A, B, C, D, E and Z.

A	B	C	D	E	Z

To work out how many rows we need, do a quick calculation: 2 to the power of our inputs. So, for this example, it would be 2^3 ($2 \times 2 \times 2$), as we have 3 inputs going into the system. This means we should have 8 rows in our table. We can now fill in all possible options. As there is an OR gate, you will have the 4 options that will exist within that truth table – but this needs to be doubled, as there will be these 4 options for when C is on, and again for when C is off.

A	B	C	D	E	Z
0	0	0			
0	0	1			
0	1	0			
0	1	1			
1	0	0			
1	0	1			
1	1	0			
1	1	1			

2 We now have to break this down into several stages. Our input A goes through a NOT gate and turns into the output named D. This is what to work out first, ignoring everything else. As the signal is going through a NOT gate, it will be the inversion of A.

A			D		
0			1		
0			1		
0			1		
0			1		
1			0		
1			0		
1			0		
1			0		

contd

3 After this, we can work out what is happening in row E. E is the output from the OR gate, which has the inputs of B and D. These are now the only rows we need to concern ourselves with. As the input is going through an OR gate, for E to be 1, B **or** D will have to be 1.

B		D	E	
0		1	1	
0		1	1	
1		1	1	
1		1	1	
0		0	0	
0		0	0	
1		0	1	
1		0	1	

4 From this, we can now work out the final column. Z is the output from the AND gate, which has input paths of E and C. This means that E **and** C need to be 1 for Z to be 1.

		C		E	Z
		0		1	0
		1		1	1
		0		1	0
		1		1	1
		0		0	0
		1		0	0
		0		1	0
		1		1	1

We now have our finished logic table.

A	B	C	D	E	Z
0	0	0	1	1	0
0	0	1	1	1	1
0	1	0	1	1	0
0	1	1	1	1	1
1	0	0	0	0	0
1	0	1	0	0	0
1	1	0	0	1	0
1	1	1	0	1	1

ONLINE TEST

Head to www.brightredbooks.net to test yourself on truth tables.

THINGS TO DO AND THINK ABOUT

Work out the Boolean expressions from the circuits below, then draw the truth table.

(a)

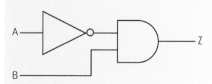

(b)

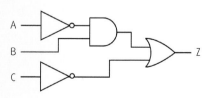

(c)

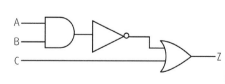

DIGITAL ELECTRONICS: TRUTH TABLES 2

CREATING BOOLEAN EXPRESSIONS AND LOGIC DIAGRAMS FROM TRUTH TABLES

Within electronic engineering, it is normal to be asked to design a logic diagram from a given truth table. To do this, focus on the combinations which give a logic 1 condition in the output column.

EXAMPLE:

A	B	C	Z
0	0	0	0
0	0	1	0
0	1	0	1
0	1	1	0
1	0	0	0
1	0	1	0
1	1	0	0
1	1	1	1

Within this circuit, there would be two occasions when the output would be 1. These are the only ones we need to consider.

A	B	C	Z
0	1	0	1
1	1	1	1

The conditions which the circuit must follow for Z to go on are:

A is at logic level 0 **and** B is at logic level 1 **and** C is at logic level 0

$(\bar{A}.B.\bar{C})$

or

A is at logic level 1 **and** B is at logic level 1 **and** C is at logic level 1

$(A.B.C)$

This means our Boolean expression would be: $Z = (\bar{A}.B.\bar{C}) + (A.B.C)$

To draw the circuit, it is always a good idea to draw vertical lines as a starting point, one for each circuit input that exists.

1

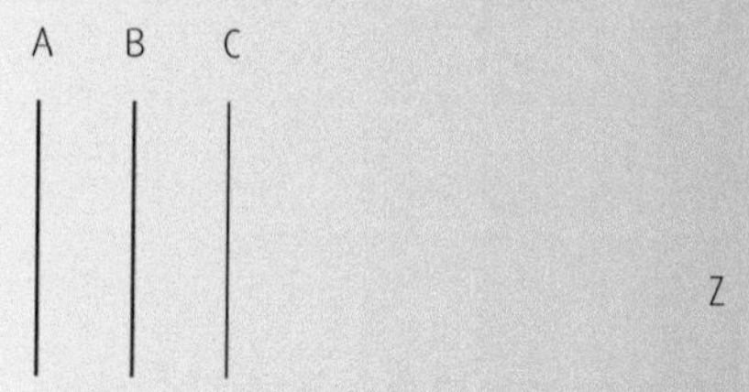

Now we break the Boolean expression down. The first part of the equation is $(\bar{A}.B.C)$. We know that it is NOT A and NOT C, so we can connect these.

2

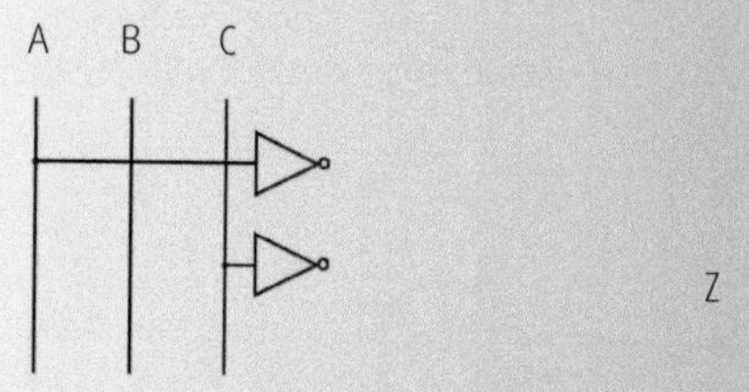

contd

We know the three inputs get AND-ed together with the B. To do this, we can use a 3-input AND gate.

3

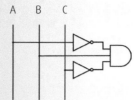

Z

We can now repeat this for the second half of the equation (A.B.C).

4

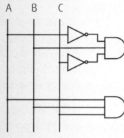

Z

To finish off the diagram, we now have to add in the OR diagram of the equation.
$Z = (\bar{A}.B.\bar{C}) + (A.B.C)$

5

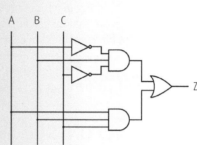

Z

⚙ ACTIVITY:

Work out the Boolean expression and the circuit diagram for the truth tables below.

(a)

A	B	C	Z
0	0	0	0
0	0	1	0
0	1	0	1
0	1	1	1
1	0	0	0
1	0	1	0
1	1	0	0
1	1	1	0

(b)

A	B	C	Z
0	0	0	1
0	0	1	0
0	1	0	0
0	1	1	0
1	0	0	0
1	0	1	1
1	1	0	0
1	1	1	1

(c)

A	B	C	Z
0	0	0	0
0	0	1	1
0	1	0	0
0	1	1	0
1	0	0	1
1	0	1	0
1	1	0	1
1	1	1	0

DON'T FORGET

Break it down into small chunks. Don't let yourself be overwhelmed!

VIDEO LINK

Get a greater understanding of converting truth tables to Boolean expressions by watching the clip at www.brightredbooks.net

 THINGS TO DO AND THINK ABOUT

Make up several Boolean circuit diagrams and truth tables, then try to figure out their corresponding Boolean expression, truth table and circuit diagram. Bring these into class and discuss them with your teacher and classmates. Discuss whether you have approached this in the correct way and got the correct answers. Make a deliberate mistake in one to see if your friends notice it!

ONLINE TEST

Head to www.brightredbooks.net to test yourself on truth tables.

PROGRAMMABLE CONTROL: MICROCONTROLLERS

WHAT IS PROGRAMMABLE CONTROL?

Programmable control is one of the activities that would concern an electronics engineer. It is about programming something called a microcontroller to do a variety of jobs that are needed in the world today. Depending on its programming, a microcontroller can be controlling anything from simple electronic toys to complex manufacturing processes within industry.

The use of microcontrollers has increased dramatically in recent times, and they are now used in almost every piece of electronic equipment, including a lot of household and everyday items such as washing machines, microwave ovens and your smartphone. In the industrial world, they are mainly used in the automation of machines, process control and conveyor lines, which has resulted in cost savings for companies, and increases in production, but also an increase in the consistency of quality. Modern consumers have come to expect a high standard of quality in the manufactured goods they buy and use, but to an engineer these are the challenges that make the profession interesting.

WHAT IS A MICROCONTROLLER?

A microcontroller is basically a very small CPU, and because of this it is often described as a 'computer on a chip'. Microcontrollers have a controller and memory all built into a single chip and can be used for sensing inputs from the real world, and then controlling output devices based on this. They are small and inexpensive, which means they can easily be built into other devices to make these products more intelligent and easier to use.

Microcontrollers are usually programmed for a specific electronic product where a specific input and/or output is pre-set – for instance, a remote-control car may use a single microcontroller to process the information for all its electronic procedures and then move its wheels depending on the input.

By altering the microcontroller program, the same 'brand' of chip can do many different tasks. In schools, PICAXE and Arduino microcontrollers are commonly used, but you may be using another variation.

Advantages of using a microcontroller

Why are microcontrollers used? There are many reasons, with the main ones being:

- **A reduced quantity of stock**: microcontrollers can be programmed to complete many different functions, so one microcontroller can do the job of many other parts. This means that companies only need to buy microcontrollers instead of all these different other components

- **Circuits will have an increased reliability**: as the microcontroller could be used to replace many different parts if it was a hardwired electronic circuit, there are fewer parts that could break.

- **A simplified product assembly and a smaller end product**: as the microcontroller can do many jobs, less complex circuitry is needed, which only takes up little space as well as making it a lot easier to build.

- **Greater product flexibility and adaptability:** as the functionality of the circuit is programmed onto the microcontroller and then inserted into the circuit, the same microcontroller can be used for many different functions. It can even be reprogrammed and put into a different circuit to complete the function of that one.

contd

- **Allows for rapid product changes:** as the nature of the circuit is dependent on what the program on the microcontroller is, it can be easily updated to be more efficient or to complete new tasks.

- **Reusable:** as the microcontroller can be updated, an old product wouldn't necessarily need to be disposed of; instead the microcontroller could just be updated and reprogrammed.

Disadvantages of using a microcontroller

- **Access to a computer:** to allow for a microcontroller to be programmed, access to a computer with the correct hardware and software is needed.

- **Costs:** the cost of the software and hardware, dependent on what type of microcontroller it is, may be expensive. The microcontroller is also expensive in comparison to other ICs.

- **Software must be up to date:** if the software is not the correct version needed for the microcontroller, it may not allow it to update. The software may also need updated frequently.

PROGRAMMING IN THE CLASSROOM

There are multiple types of microcontrollers that could be used within your school, and each will work with a different programming language. You may have used one already in your learning, such as the BBC MicroBit to program some basic programs and systems. Within this course, you will have to learn a high-level language, and the hardware you use does not matter. For example, you may continue to use the BBC MicroBit, or you may use Arduino, which is based on the computing language C++, or use a STAMP controller that uses the PBASIC language. Any high-level language is acceptable within the National 5 Engineering Science Course.

Only the internal assessments and the final assignment will assess the computer language you use, so do not feel you have to remember how to write it at this stage.

THINGS TO DO AND THINK ABOUT

Look around your home and try to identify products that will be controlled by a microcontroller. Write down the reason why you think each has a microcontroller, and discuss this in class with your teacher and classmates.

DON'T FORGET

No matter what microcontroller you use, the all work in the same basic way. You are at no advantage or disadvantage over other schools because of your kit.

ONLINE

Go to www.brightredbooks.net to get more information on microcontrollers.

DON'T FORGET

You do not need to remember these, as this section is not needed for the course, but it is good to have an understanding of how a microcontroller works.

ONLINE TEST

Head to www.brightredbooks.net to test yourself on microcontrollers.

PROGRAMMABLE CONTROL: FLOWCHARTS 1

FLOWCHARTS

Before a computer program is written, you have to work out what is happening in it by drawing a flowchart. As flowcharts are drawn graphically, they often make programs easier to understand.

Flowcharts use several symbols within them which represent different parts of a program.

Start/Stop	**Start/stop symbol** The start/stop symbol is a rectangular shape with rounded ends. Each flowchart must contain one start and sometimes a stop symbol, depending on whether the program is to stop or is to be a continuous loop that will never end. If it is found within a sub-procedure, the end should say 'return', as this will then return it to the main program rather than stop it.
Wait x Seconds	**Wait symbol** The 'wait' symbol is a rectangle. The text inside the symbol explains how long a command will be delayed for.
Pin x HIGH	**Output symbol** The 'output' symbol is a parallelogram. The text inside this should explain which output pins are switched on or off at any time. If pins are known, then the actual pin number should be mentioned. If a pin is on, it is known to be HIGH; if it is off, it is known to be LOW.
Repeated x Times? No Yes	**Decision box** The decision box is a diamond. This is used to check whether something has been completed. Sub-procedure
Name of Procedure	**Sub-procedure** A sub-procedure is a small section of code that can be called upon but is not within the main program. After the sub-procedure is finished, it should then return to the main program.

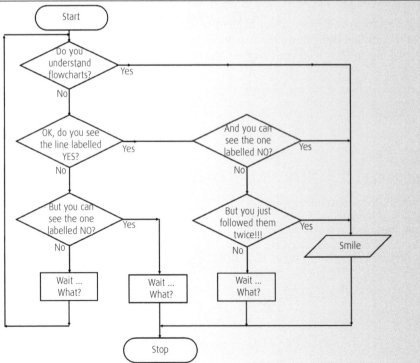

DON'T FORGET

Make sure you take your time to understand flowcharts and know how to write and read them. You will need to do one for your internal assessment and your final assignment, and you could be asked to draw one in your exam!

CONTINUOUS LOOPS

Quite frequently, it is necessary to create programs that have an infinite loop. This means the program will loop forever and never stop. To achieve this, we do not use the 'stop' symbol – instead, we draw a feedback loop going back to the beginning. This is the most common type of programming, as it is very rare that you would want your code to end when it has been completed. This would shut everything down, meaning a programming engineer would need to come to the system to reset it or to reprogram it for the system to work again. A good example would be traffic lights: you would not want the programme controlling them to end as soon as it has gone through the light combinations only once each! You want it to last forever, or else it could cause a lot of trouble on the roads for all vehicles and pedestrians.

EXAMPLE:

A set of temporary traffic lights is required for a system of roadworks. The operation of the system is shown below.

- The lights should go red for 12 seconds
- The amber will then come on for 5 seconds
- The red and amber lights will switch off, and the green will switch on for 12 seconds
- The green switches off, and the amber goes on for 4 seconds, then switches off
- The system repeats.

Input connection	Pin	Output connection
	2	Red light
	1	Amber light
	0	Green light

Draw the flowchart.

THINGS TO DO AND THINK ABOUT

Create a flowchart for a dispensing machine making a hot drink. Does it have sugar? Does it have milk? Remember you have to tell the system to stop these commands as well as starting them.

DON'T FORGET

Yenka can build flowcharts; and, if used alongside a microcontroller, electronic inputs and outputs can be connected to simulate the flowchart.

DON'T FORGET

When creating a flowchart, you should **always** refer to the pin number and not to the command.

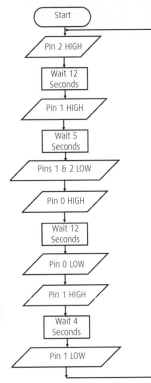

Flowchart:
Start → Pin 2 HIGH → Wait 12 Seconds → Pin 1 HIGH → Wait 5 Seconds → Pins 1 & 2 LOW → Pin 0 HIGH → Wait 12 Seconds → Pin 0 LOW → Pin 1 HIGH → Wait 4 Seconds → Pin 1 LOW

PROGRAMMABLE CONTROL: FLOWCHARTS 2

FIXED LOOPS IN A FLOWCHART

Some flowcharts also have something known as a fixed loop. This is where a given task is expected to complete a set amount of times before the program ends.

EXAMPLE:

A toy robot is controlled by a microcontroller and follows the sequence shown below.

1. A start switch must be activated for it to switch on
2. The legs start moving
3. After 2 seconds, the LEDs on the gun light up
4. After 2 seconds, the arms move
5. After 5 seconds, the arms stop and the LEDs go off
6. The legs then stop moving
7. Steps 2–6 will repeat three times
8. The sequence will then repeat.

Input connection	Pin	Output connection
	3	Legs
	2	Arms
	1	LEDs
Start switch	0	

Draw the flowchart.

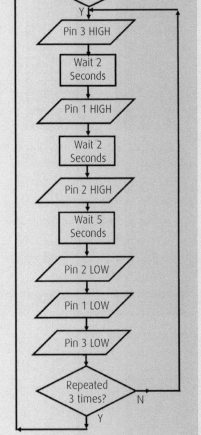

DON'T FORGET ➕

If you find yourself getting confused by having to name the pins, write little notes beside to explain what it is your flowchart is doing.

⚙️ **ACTIVITY:**

(a) A car-park barrier uses a microcontroller to operate. The operation of the system is shown below.

- When the button is pressed, the machine will eject the parking ticket.
- Once the ticket is taken, it will raise the barrier for 5 seconds.
- When it has sensed the car has moved away, it will lower the barrier for 5 seconds.
- The sequence will repeat.

Complete the flowchart for the barrier operation.

contd

(b) A water jet uses high-pressure water to cut metals, such as sheet steel.

The position of the water jet is controlled by motors A and B, which are operated by a microcontroller. The input and output connections to the microcontroller are shown in the table.

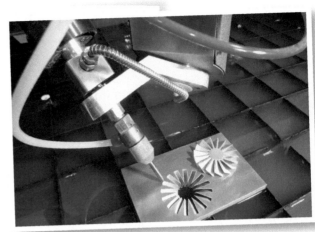

Input connection	Pin	Output connection
	7	Motor A Forward (right)
	6	Motor A Reverse (left)
	5	Motor B Forward (forward)
	4	Motor B Reverse (back)
	3	Water Jet On
	2	
	1	
	0	

The cutter is required to perform the following sequence of operations:

1. Switch on the water jet
2. Move right for 1·5 seconds
3. Move forward for 3·5 seconds
4. Repeat steps 2 and 3 five times
5. Switch off the water jet and the motors.

Draw the flowchart for this sequence. The flowchart **must** include appropriate pin numbers.

THINGS TO DO AND THINK ABOUT

Design part of a sequence to control traffic lights by drawing a flowchart. When a button is pressed, a delay of 10 seconds will occur, then it will switch off a green light and turn on a red one. Then, 10 seconds after this, it will switch off the red and switch on an amber light. This will go on for a second and off for a second, repeating 10 times. After this, the green light will go on. The sequence will then repeat.

DON'T FORGET

Make sure you always read the question properly. In the exam, it is likely that the question will ask you to refer to the pins, which means that if you do not, you will lose marks!

ONLINE

Follow the link at www.brightredbooks.net to help you get a greater understanding of flowcharts.

ONLINE TEST

Head to www.brightredbooks.net to test yourself on flowcharts and continuous loops.

PROGRAMMABLE CONTROL: ARDUINO 1

VIDEO LINK

Watch the video at www.brightredbooks.net to see the importance of programming.

DON'T FORGET

You will not be asked to write code in the exam. However, if you are asked to develop a flowchart, you will be asked to refer to the input and output pin numbers.

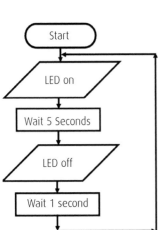

WRITING CODE

Once a flowchart has been drawn, it is necessary to convert it into something the microcontroller can read, and this is done using a high-level computer language. The program created can vary in language. The one you learn will depend on what resources are available in your school. Although it will differ from school to school, it likely you will learn one of two languages. The first is a variation of a language known as C. You will learn this if your school uses Arduino. The second is PBASIC, which you will be learning if your school has STAMP or PICAXE boards. It doesn't matter which language you learn, as any high-level language is acceptable.

The following pages will be broken into two possible languages and hardware: C using Arduino; and PBASIC.

ARDUINO

When you load up your Arduino sketch (the name Arduino calls a program), it may or may not have the following already written below. It completely depends on which version of Arduino software has been installed.

```
void setup()
{
}
void loop()
{
}
```

- **Void setup()** commands are ones that the microcontroller will only read once. This is where you tell your board which pins are to be treated as inputs, and which ones are to be treated as outputs.

- **Void loop()** commands are ones that will run continuously in a loop forever while the Arduino is on. This is where your main program should be written.

When writing your sketch, the commands need to be between braces **{ }**, as this tells the microcontroller that everything it needs to know about this section will be in between these.

We are going to learn how to write the code for this basic flowchart.

The first thing that has to be done is to name the sketch. This should be **above** the void setup() and void loop() sections. This can be any name whatsoever, but it must start with a /* and end with a */.

Your next step is to name the pins. This makes it easier to understand a program, as we can refer to it by a name rather than constantly referring to the pin number, which could get confusing. For example, if we are going to connect a red LED to pin 12, we could create an integer and call the pin redLED like this:

int redLED = 12;

This should be placed in between the name of the program and void setup.

We now need to work within the 'void setup ()' section. This is where we tell the Arduino board if the pins we are using will be acting as an input or as an output. To do this, we need to use the 'pinMode' command and set our LED as an output. This is because an LED is an output device.

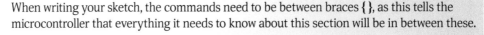

contd

```
/* My first Arduino sketch */

int redLED = 12;

void setup()
{
pinMode(redLED, OUTPUT);
}
```

As you can see when writing the sketch, your commands collectively need to be between braces { }, as this tells the microcontroller that everything it needs to know about this section will be in between these.

You will also notice that in the command 'pinMode' the M is a capital letter – this is because Arduino uses something called camelCase. This is where it would normally be two or more words, but, to save memory, the second and any following words start with a capital letter instead.

At the end of every command, you will also notice that there is a semi-colon (;). This is to tell the Arduino board that this is the end of a line and that it should move onto the next one.

The actual code has to be written in the void loop() section:

```
void loop()
{
  digitalWrite(redLED, HIGH);
  delay(5000);
}

void loop()
{
  digitalWrite(redLED, HIGH);
  delay(5000);
  digitalWrite(redLED, LOW);
  delay(1000);
}
```

To write a command, we use the 'digitalWrite' command, and state if the signal is HIGH (logic 1) or LOW (logic 0). To create a pause, we use the 'delay' command. When creating a delay, the time is measured in milliseconds (thousandths of a second) e.g. second will be 1000.

You will now have a finished sketch like this.

```
/* My first Arduino sketch */

int redLED = 12;

void setup()
{
pinMode(redLED, OUTPUT);
}

void loop()
{
  digitalWrite(redLED, HIGH);
  delay(5000);
  digitalWrite(redLED, LOW);
  delay(1000);
}
```

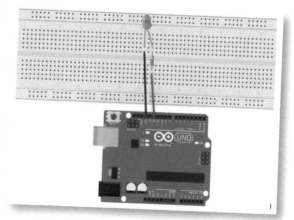

Before we test the sketch, we have to connect the LED to the breadboard, using skills you have learned within the classroom during the Analogue Electronics section of the course.

By building the circuit as shown, and sending the microcontroller the code that has just been created, it will complete the tasks in the flowchart.

THINGS TO DO AND THINK ABOUT

Sign up to 123dcircuits.io to build and test this circuit. Experiment with the delay, and see what happens.

DON'T FORGET

It is extremely important when writing a sketch that the syntax and spelling are exactly correct, or the program will not work. For example, redLED is not the same as REDLED.

VIDEO LINK

Watch the video on how to connect a breadboard at www.brightredbooks.net

DON'T FORGET

Make sure you also connect the LED properly. The long leg (anode) should be on the + side, and the short leg (cathode) should be to the ground.

ONLINE TEST

Test yourself on Arduino at www.brightredbooks.net

PROGRAMMABLE CONTROL: ARDUINO 2

UNDERSTANDING THE SKETCH

As you may have discovered, when writing a program using a high-level computer language, it can be confusing at times reading each line and then knowing what it is supposed to do – and it will only get more confusing as you create larger programs. To solve this problem, we can use // at the end of a line of programming. By using // it tells the microcontroller to stop reading and move onto the next line.

For example, instead of just having ...

```
void loop()
{
  digitalWrite(redledPin, HIGH);
  delay(10000);
  digitalWrite(redledPin, LOW);
}
```

we could write:

```
void loop()
{
  digitalWrite(redledPin, HIGH);    // Red LED switched on
  delay(10000);                     // pause 10 seconds
  digitalWrite(redledPin, LOW);     // Red LED switched off
}
```

This lets us know that the LED will be switched on and this command will stay the same for 10 seconds. After this, the LED will be switched off.

ERROR-CHECKING

If you have already made a mistake, for example in your syntax, the Arduino program will tell you something is wrong and will even tell you where the mistake is by highlighting it. In the example above, it says:

Expected ';' before digitalWrite

This informs you that there is no ; at the end of the previous line, and it has highlighted the line where it found the error.

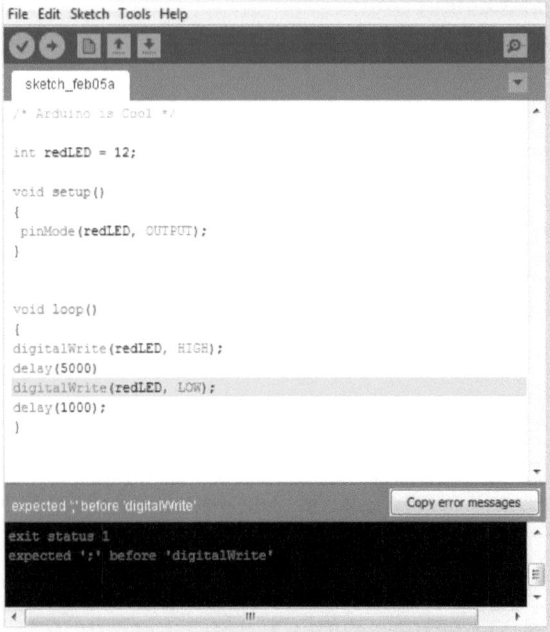

```
File Edit Sketch Tools Help

sketch_feb05a

/* Arduino is Cool */

int redLED = 12;

void setup()
{
  pinMode(redLED, OUTPUT);
}

void loop()
{
digitalWrite(redLED, HIGH);
delay(5000)
digitalWrite(redLED, LOW);
delay(1000);
}

expected ';' before 'digitalWrite'              Copy error messages
exit status 1
expected ';' before 'digitalWrite'
```

DECISIONS

If there is a decision box that is asking a question in our flowchart, we have to use the 'if...else' command.

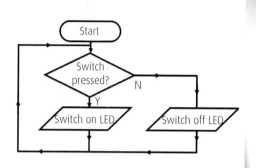

This is used when there should be an option of outputs that is dependent on what the input is. For example, if a switch is pressed, a light will go on; otherwise, the light will stay off.

When writing the code, we use the **digitalRead** command. This means that Arduino will check the input for a digital command – and **if** the desired input has been achieved, it will complete that output. Otherwise (**else**), it will complete the other option. So, the command that would be used within the code for this flowchart would be:

```
if (digitalRead(switchPin) == HIGH)
   {
   ...
   }
```

You will notice that there is no ; after the **if** command. This is because you will be opening up a new set of braces. Within these braces, you will state what will now happen when the switch pin is pressed.

You may also notice that the command uses two equals signs: (**switchPin**) == HIGH). This is because it is **checking** to see **if** the pin is high. If we said (**switchPin**) = HIGH), this instead would be **telling** the computer that the pin **is** high rather than asking to check it.

DON'T FORGET

When using a 4-pin switch such as this, it is a good idea to include a drop-down resistor. This will ensure that the logic level of the switch is always at 0 unless the button is pressed.

EXAMPLE:

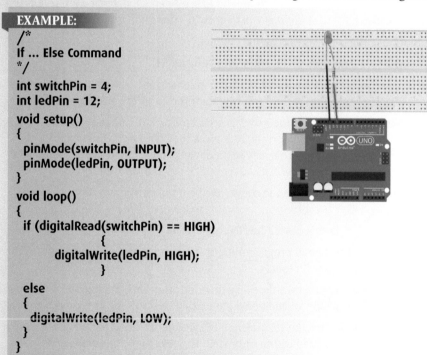

```
/*
If ... Else Command
*/
int switchPin = 4;
int ledPin = 12;
void setup()
{
  pinMode(switchPin, INPUT);
  pinMode(ledPin, OUTPUT);
}
void loop()
{
  if (digitalRead(switchPin) == HIGH)
        {
        digitalWrite(ledPin, HIGH);
            }
  else
  {
    digitalWrite(ledPin, LOW);
  }
}
```

ONLINE

Read more about 'if...else' loops at www.brightredbooks.net

ONLINE TEST

Head to www.brightredbooks.net to test yourself on Arduino.

Within this program, it will switch the LED on when the button is pressed, and switch off when it is not.

THINGS TO DO AND THINK ABOUT

A circuit is needed to be designed for a queuing system for a theme park. They want a system where, when one switch is pressed, a green LED lights up telling customers to walk onto the ride, but when another switch is pressed, a red LED lights up telling customers to stop.

Create the flowchart for this system, and then design the code. Build and test your solution using 123dcircuits.io

PROGRAMMABLE CONTROL: ARDUINO 3

CREATING A COUNTER

It is often useful to repeat the same part of a program a number of times, for instance when flashing an LED. Instead of constantly pressing a button, we could press it once and it would follow a sequence to flash a certain amount of times.

For example, if it were to follow this flowchart, it would flash an LED 10 times when the button is pressed.

To do this, we have to create the sketch using the 'if...else' command that you have learned previously to check if the switch is pressed first.

void loop()
{
  if (digitalRead(switchPin) == HIGH)
}

This 'reads' your circuit to see if the button has been pressed, or in other words, 'if' switch pin = 1 (HIGH).

Later, we will add the 'else' part of this command to see what will happen if it has not been pressed, but we have to state what will happen 'if' the switch is pressed first.

Our next block of code is for setting up the counter.

{ for (int counter = 1 ; counter <= 10; counter = counter +1)

There are three separate statements within this command that are each separated by a semi-colon. The first statement is the initialisation of the counter variable.

int counter = 1:	This tells the Arduino board to start at 1 when it is counting. This is important, as it gives the board a starting point for the counter. This may seem obvious, but a computer does not have common sense like we do – it only knows **exactly** what we tell it.

The next part of the command tells the computer the number of times it has to repeat this process.

counter <= 10:	In this case, it has to count to 10.

The final command in this line of programming tells the computer how much to add to the value of the pin each time it repeats.

counter = counter +1:	In other words, every time the program completes a loop, it has to add an additional loop until it reaches the previously set limit.

After this command is written, the 10x flashing LED aspect of the program can be written. This has to be contained within { } braces. The reason for this is that the microcontroller reads the program one line at a time. Without the braces to tell it that this is a block of code to be repeated, it will only repeat the next line.

void loop()
{
   if (digitalRead(switchPin) == HIGH)
      { for (int counter = 1 ; counter <= 10; counter = counter ++)
        { digitalWrite(ledPin, HIGH);
        delay (250);
        digitalWrite(ledPin, LOW);
        delay (250);
        }
      }

contd

We now have to add the 'else' part of the command from our decision box. According to our flowchart, this switches off the LED and loops back to the beginning.

```
void loop()
{
   if (digitalRead(switchPin) == HIGH)
      { for (int counter = 1 ; counter <= 10; counter = counter +1)
      { digitalWrite(ledPin, HIGH);
         delay (250);
         digitalWrite(ledPin, LOW);
         delay (250);}
   }
else
   {
   digitalWrite(ledPin, LOW);
   }
}
```

DON'T FORGET

If you start a bracket or a brace within a program, make sure you finish it, or it will not run. Within the Arduino software, you can put your cursor over a bracket or brace, and it will highlight the corresponding one.

ONLINE

Visit www.brightredbooks.net to help your understanding of the 'for' loop and counters.

LABELS

Another way we could have written this sketch is by using labels. This breaks the program into branches so that it is easier to understand.

```
/*
Counter Sketch with Labels
*/

int ledPin = 12;
int switchPin = 4;

void setup()
{
   pinMode(ledPin, OUTPUT);
   pinMode(switchPin, INPUT);
}

void loop()
{
MAIN_LABEL:    // This is a label I can now refer to in my program
   if (digitalRead(switchPin) == LOW)
      {
      goto MAIN_LABEL;   // This will create a loop that will 'goto' the
      }     // label I've created until the input condition
   // changes
      for (int counter = 1 ; counter <= 10; counter = counter +1)
         { digitalWrite(ledPin, HIGH);
         delay (250);
         digitalWrite(ledPin, LOW);
         delay (250);
         }

goto MAIN_LABEL;
}
```

Creating labels is a way in which we can create points in the program to loop back to. Using the 'goto' command will tell our program where to 'go to', as can be seen above. In this case, if it senses the pin is High, it will continue reading through the program and automatically move onto the next line of code, looping it back to the 'MAIN_LABEL'. This is a way to simplify certain programs and make them easier to understand.

THINGS TO DO AND THINK ABOUT

Using the same breadboard set-up for the decision box, upload the counter program to the Arduino software or 123Dcircuits.io and test the program. Does it work as you thought it should? Change the code so that you are including labels. Does this work as you expected?

ONLINE TEST

Head to www.brightredbooks.net to test yourself on Arduino.

PROGRAMMABLE CONTROL: ARDUINO 4

MOTOR CONTROL

We can connect any output device to the Arduino board and can then program it to do whatever we want. So far, we have only used LEDs, switching them on and off. This could have been **any** output device, but it would still have only been on or off.

When dealing with a motor, there are multiple ways of controlling it, from using a shield to a driver chip. This will purely depend on what resources your school has. The most common way is by using a type of IC chip called a 'motor driver IC chip'. This is built up of many internal transistors to create something called an H-bridge. An H-bridge is an electronic circuit that enables a voltage to be applied across a load in either direction. These circuits are often used in robotics and other applications to allow DC motors to run forwards and backwards. In essence, it acts like a DPDT switch.

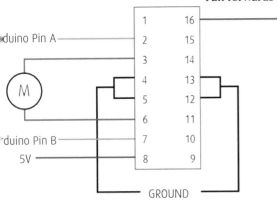

In this case, we will use an 'SN754410 motor driver IC chip' to drive the motor, but it may be another chip being used. Your teacher will show you the correct way to connect your chip.

To control the motor, the chip needs to be connected to a 5 V power supply and the ground. To connect this, you find the indent or semi-circle that will be on the side of the chip. This then faces left when connecting to your breadboard. Your 5 V supply then goes into the top left and bottom right pins, and your 0 V goes into the middle pins.

It then needs two pins to control the motor. When one pin is high, it will turn clockwise, and when the other pin is high, it will turn anti-clockwise.

EXAMPLE:

If I wanted a simple program to turn the motor clockwise for 5 seconds, stop for 1 second, turn anti-clockwise for 5 seconds, then stop for 3 seconds, I would run the following program.

```
/*
Motor driver control
*/
int motor1APin = 6;      // H-bridge leg 1
int motor2APin = 7;      // H-bridge leg 2
void setup()
{
pinMode(motor1APin, OUTPUT);
pinMode(motor2APin, OUTPUT);
}
void loop()
{
digitalWrite(motor1APin, LOW);      // set leg 1 of the H-bridge low
digitalWrite(motor2APin, HIGH);     // set leg 2 of the H-bridge high
delay (5000);

digitalWrite(motor1APin, LOW);      // set leg 1 of the H-bridge low
digitalWrite(motor2APin, LOW);      // set leg 2 of the H-bridge low
delay (1000);
digitalWrite(motor1APin, HIGH);     // set leg 1 of the H-bridge high
digitalWrite(motor2APin, LOW);      // set leg 2 of the H-bridge low
delay (5000);
digitalWrite(motor1APin, LOW);      // set leg 1 of the H-bridge low
digitalWrite(motor2APin, LOW);      // set leg 2 of the H-bridge low
delay (3000);
}
```

contd

⚙ ACTIVITY:

Design a flowchart and program for a remote-control car so that when one button is pressed it will drive forwards, and when another button is pressed it will drive backwards.

Upload this to the Arduino software or 123Dcircuits.io to test if it works.

THE 'WHILE' COMMAND

Our program for the remote-control car still does not work as well as it could do. This is because the system only works when we have our finger on the button – as soon as it's released, it stops that command.

The way to resolve this is by using the 'while' command. This creates a loop within the program that will loop continuously, and infinitely, until the command inside its brackets () becomes false.

For example, within my program, I may have set up a command that will keep it driving forwards until the reverse (or an additional stop) is pressed. The '!' is used to invert the signal. This means it will continue to do this while the reverse switch is **not** pressed.

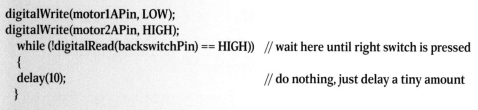

```
digitalWrite(motor1APin, LOW);
digitalWrite(motor2APin, HIGH);
  while (!digitalRead(backswitchPin) == HIGH))   // wait here until right switch is pressed
  {
  delay(10);                                      // do nothing, just delay a tiny amount
  }
```

You will notice there is also a delay of 10ms inserted after this – this command does nothing except create a slight pause between it changing directions with a new command. This is good practice to prevent it from damaging the motor.

 ONLINE

Develop your understanding of 'while' loops by following the link at www.brightredbooks.net

ⓘ THINGS TO DO AND THINK ABOUT

A local theme park is creating a new rollercoaster called the 'King Cobra'. To help advertise this, they want large interactive models of cobras to be placed in local towns and villages. These models will have glowing eyes and a tongue that will stick out 3 times whenever someone stands near it. To achieve this, a pressure pad will be in front of it that will send a high signal to a microcontroller whenever a person stands on it. The microcontroller must then operate a motor forwards then backwards that will be attached to the gearing system for the tongue.

Draw the flowchart and create the program that would complete this task.

✓ **ONLINE TEST**

Head to www.brightredbooks.net to test yourself on Arduino.

PROGRAMMABLE CONTROL: PBASIC 1

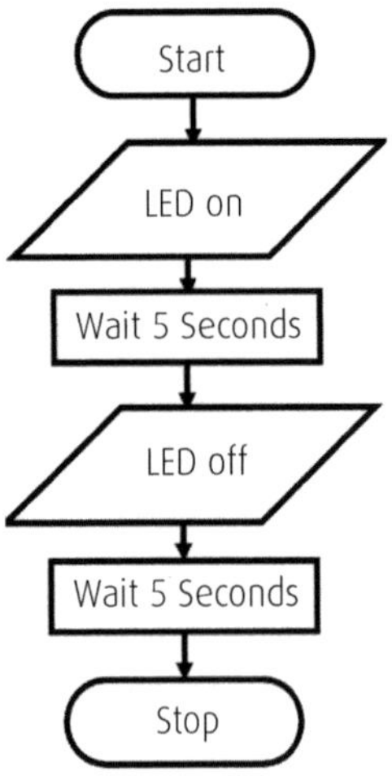

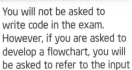

USING PBASIC

To learn how to use PBASIC, we are going to write a small program for the flowchart here.

If you have read through the Arduino section of this chapter, you may notice that this flowchart is slightly different from the Arduino one, as this ends. Within Arduino, it recognises that 99 per cent of all programs should never stop after they have completed, so it automatically creates a loop. PBASIC does not do that. We will cover how to create loops in PBASIC later in this book.

First, we have to tell the microcontroller which pins are outputs and which pins are inputs. We do this with the 'let dirs' command. The board will have 8 pins that range from 0 to 7, and is read from right to left.

For example, if we wanted pins 0–3 to be inputs and pins 4–7 to be outputs, we would state:

let dirs = %11110000

with 1 being an output and 0 being an input.

After this, we have to create a label for the program. This is always required, and it doesn't matter what you call it – it makes sense, though, to call it something suitable or related to the program.

When switching the inputs or outputs on or off, we refer to the state of the signal and whether it is HIGH (logic 1) or LOW (logic 0).

To create a delay in the program, we use the 'pause' command. When creating a pause, we measure the time in milliseconds (1000ths of a second).

e.g. 1 second = 1000
     5 seconds = 5000

let dirs = %11110000

main:
    high 7
    pause 5000
    low 7
    pause 1000

    end

DON'T FORGET

You will not be asked to write code in the exam. However, if you are asked to develop a flowchart, you will be asked to refer to the input and output pin numbers.

ONLINE

Follow the link at www.brightredbooks.net to download PICAXE software, allowing you to simulate programs.

UNDERSTANDING THE SKETCH

As the program becomes more complicated, it is essential we understand what it is doing. It can become confusing reading each line and then knowing what is supposed to happen.

To solve this problem, we can use an apostrophe at the end of a line of programming and then type in an explanation in clear English. By using an apostrophe, it tells the microcontroller to stop reading and move onto the next line.

For example, instead of just having …

let dirs = %11110000
main:
    high 7
    pause 5000
    low 7
    pause 1000

    end

DON'T FORGET

Use the tab key on the keyboard to create large even spaces.

contd

we could write:

```
let dirs = %11110000
main:
    high 7          ' pin 7 switches on
    pause 5000      ' pauses for 5 seconds
    low 7           ' pin 7 switches off
    pause 1000      ' pauses for 1 second

    end             ' end of program
```

LABELS

A label, for example 'main:' in the program above, can be just about any word. When the label is first defined, it must end with a colon (:). The colon 'tells' the computer that the word is a new label. Creating labels is a way in which we can create points in the program to loop back to by using the 'goto' command. This tells the program to 'go to' a certain place.

For example, if I wanted the program to work in a continuous loop instead of ending, I could remove the 'end' command and use the 'goto' instead.

```
let dirs = %11110000
main:
    high 7          ' pin 7 switches on
    pause 5000      ' pauses for 5 seconds
    low 7           ' pin 7 switches off
    pause 1000      ' pauses for 1 second

goto main           ' loops the program back to the main label
```

DON'T FORGET

It is extremely important when writing a sketch that the syntax and spelling are exactly correct, or the program will not work. For example, redLED is not the same as REDLED.

SYMBOLS

Another thing that can be done to help you understand your sketch is to use symbols. This renames the pin so you can make it more suitable for its purpose.

In addition to the program, I could rename pin 7 'redLED' so I know exactly what it does.

```
let dirs = %11110000

Symbol redLED = 7      ' define pin with the name redLED

main:
    high redLED        ' pin 7 switches on
    pause 5000         ' pauses for 5 seconds
    low redLED         ' pin 7 switches off
    pause 1000         ' pauses for 1 second

goto main              ' loops the program back to the main label
```

ONLINE TEST

Head to www.brightredbooks.net to test yourself on PBASIC.

THINGS TO DO AND THINK ABOUT

A set of temporary traffic lights is required for a system of roadworks. The operation of the system is shown below.

- The lights should go red for 10 seconds
- The amber will then come on for 4 seconds
- The red and amber lights will switch off, and the green will switch on for 15 seconds
- The green switches off, and the amber goes on for 5 seconds, then switches off
- The system repeats.

Draw the flowchart for this system, then try to work out the code.

PROGRAMMABLE CONTROL: PBASIC 2

DECISIONS

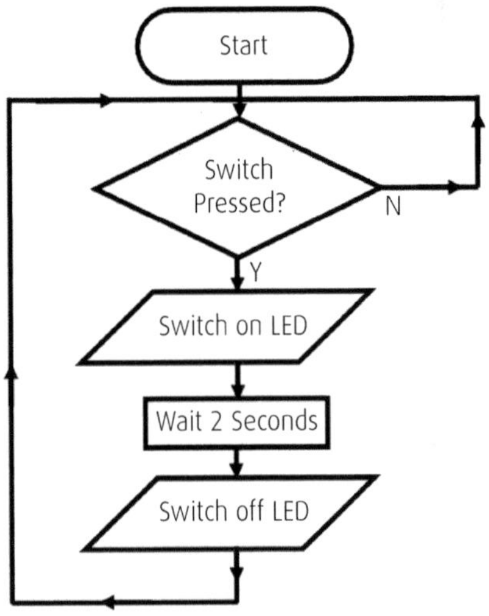

If there is a decision box asking a question in our flowchart, we have to use the 'if... then' command. This tells the computer that 'if' these conditions are met, 'then' this is what is going to happen.

For example, in the flowchart above, it is checking constantly if the switch has been pressed. Only once it senses that the pin for this switch has gone high will it move on through the program. This is achieved by using the 'if...then' command in conjunction with labels, as seen in the program below.

```
let dirs = %11110000
main:                        ' make a label called 'main'
    if pin0 =1 then flash    ' jump to the label 'flash' if the input is on
    goto main                ' else loop back around to 'main'

flash:                       ' make a label called 'flash'
    high 7                   ' switch output 7 on
    pause 2000               ' wait 2 seconds
    low 7                    ' switch output 7 off
    goto main                ' jump back to the 'main' label
```

If the switch is then pushed, the program jumps to the label called 'flash'. The program then switches pin 7 on for two seconds before returning to the main loop.

 **ACTIVITY:**

Part of a circuit is needed to be designed for a slot machine. Once a coin is sensed and the start button is pressed, a green LED will flash 3 times, being on for 0·5 seconds, then off for 0·5 seconds. When the stop button is pressed, a red LED will light up for 3 seconds. The system will then repeat the sequence.

Input connection	Pin	Output connection
	4	Red LED
	3	Green LED
Stop switch	2	
Start switch	1	
Coin sensor	0	

Draw a flowchart and write the code for the system.

DON'T FORGET

You may notice that the spelling in the 'if...then' line – 'pin0' – is all one word (without a space). You will also notice that only the label is placed after the command 'then' – no other words apart from a label are allowed to be placed here.

CREATING A COUNTER

It is often useful to repeat the same part of a program numerous times, for instance when flashing an LED. Instead of constantly pressing a button, we could press it once and it would follow a sequence to flash a certain amount of times. To do this, we can use a 'for ... next' loop.

The number of times this section of code has to be repeated is stored within the RAM memory of the stamp controller. There are 10 different variables you can use to do this, which are labelled b0 to b9. This can then be utilised with the symbol command to make it easier to read.

contd

Flowchart (left margin):

- Start
- Switch Pressed? — N (loops back)
- Y
- Switch on LED
- Wait 2 Seconds
- Switch off LED

```
Symbol redLED = 7
Symbol counter = b0

main:     for counter = 1 to 10    ' start a 'for … next' loop
          high redLED              ' switch pin 7 high
          pause 1000               ' wait for 1 second
          low redLED               ' switch pin 7 low
          pause 1000               ' wait for 1 second
          next counter             ' end of 'for … next' loop
          end                      ' end program
```

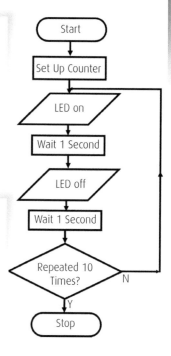

MOTOR CONTROL

We can hook any electronic component to the STAMP or PICAXE board and then program it to do whatever we want. So far, though, we have only used LEDs with variations of switching on and off. This could have been **any** output device, but it would still have only been on or off.

When dealing with a motor, there are multiple ways of controlling it, from using an output driver module to a driver chip. This will purely depend on what your school resources are like. The most common way is by using an output driver. This is built up of many internal transistors to create something called a push-pull driver. This is an electronic circuit that enables a voltage to be applied across a load in either direction. These circuits are often used in robotics and other applications to allow DC motors to run forwards and backwards. In essence, it acts as a DPDT switch by changing whether each pin is on or off, it will change the output of the motor.

PIN A	PIN B	OUTPUT
LOW	LOW	OFF
HIGH	LOW	CLOCKWISE
LOW	HIGH	ANTI-CLOCKWISE
HIGH	HIGH	OFF

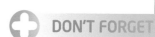

THINGS TO DO AND THINK ABOUT

Part of a circuit operating a holographic projector is to be designed for a haunted-house attraction in a theme park. When movement is sensed, a holographic projector will be switched on and it will move from side to side four times. To do this, the microcontroller must operate a motor that is attached to a gearing system forwards for 3 seconds, pause for 0·5 seconds, and then move backwards for 3 seconds. Once completed, the system will reset.

Draw the flowchart and create the program that would complete this task.

Input connection	Pin	Output connection
	3	Motor Pin B
	2	Motor Pin A
	1	Projector
Movement Sensor	0	

DON'T FORGET

If you build a flowchart and connect it to a microcontroller in Yenka, it can create your own code for you. Go to 'PIC PROGRAMMING' – 'TOOLS' – 'BASIC VIEWER'.

ONLINE TEST

Head to www.brightredbooks.net to test yourself on PBASIC.

MECHANISMS AND STRUCTURES

MECHANISMS AND DRIVE SYSTEMS: MOTION

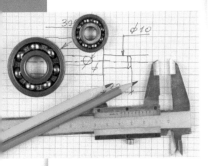

WHAT IS MECHANICAL ENGINEERING?

Mechanical engineering is a diverse discipline that looks at how objects and forces react to one another. It is a career path that looks at the design and manufacture of objects – everything from small individual parts and devices like inkjet printer nozzles, to larger systems such as spacecraft or machinery for the oil industry. Mechanical engineering looks at how these items are created, how they work and how they can be improved.

Mechanical engineers play a central role in a huge selection of different industries and potential career paths. Some examples of these can be found in the automotive industry (from chassis to the engine, transmission and sensors), the aerospace industry (aircraft engines and their control systems), biotechnology (from the design and creation of medical implants and prosthetic devices, to the use of fluidic systems in the pharmaceutical industry), the computers and electronics industry (disc drives, printers, cooling systems), energy conversion (gas and wind turbines), environmental control (air-conditioning and refrigeration systems), automation (robots and robotic control) and the manufacturing industry (machining and fabrication).

Basically, if it moves, a mechanical engineer was involved in its design and innovation.

WHAT IS A MECHANICAL SYSTEM?

A mechanism changes an input motion and force into another output motion and force.

All mechanisms must:
- involve some kind of motion
- involve some kind of force
- make a job easier to do
- need some kind of input to make them work
- produce some kind of output.

Mechanisms are commonplace in modern society – so much so that we will use complex mechanisms every day without even realising it. For example, since you have awoken this morning, you will have used several door handles and light switches. You may have even used a tap, hair straighteners or spray deodorant. When you got on the bus, your bus may have lowered to let you on. The engine of the bus then drove you to school. These are all, or contain, forms of mechanisms.

Mechanisms play a vital role in industry. While many industrial processes now involve a lot of electronic control systems, it is still mechanisms that provide the 'muscle' to do the work. For example, some form of mechanisms will provide the force needed to press sheet metal into shapes so they can be used for car body panels, or to lift large components from one place to another. It is only through the use of different mechanisms that industry can make the products you use every day.

MOTION

To understand how a mechanism works, it is vital you know the ways it can move. There are four forms of motion:

Linear motion ⟶

This is movement in a straight line, going in one direction – for example, a paper guillotine.

VIDEO LINK

Watch the videos at www.brightredbooks.net to see what it is like to be a mechanical engineer and to understand what mechanical engineering is.

VIDEO LINK

Check out the clip at www.brightredbooks.net to see a career route within mechanical engineering that you may not have considered.

DON'T FORGET

Do not mix up a mechanical engineer with a mechanic. A mechanical engineer will design the system; it is a mechanic who will install it. Do not fall into this trap when writing about the role of an engineer.

contd

60

Reciprocating motion

This is moving backwards and forwards in a straight line – for example, the needle on a sewing machine.

Rotary motion

This is turning in a circle – for example, the hands on a clock.

Oscillating motion

This is swinging backwards and forwards in an arc – for example, the movement of a rocking horse.

ONLINE

Visit www.brightredbooks.net to get a greater understanding of how this motion is used.

⚙ ACTIVITY:

What types of motion do the following activities show when they are being used or carried out? (They may contain more than one.)

(a)

(b)

(c)

⚠ THINGS TO DO AND THINK ABOUT

Look around your home to find five objects that involve some form of mechanism. Write down the types of motion they use, and whether they are used to transmit force or movement.

ONLINE TEST

Head to www.brightredbooks.net to test yourself on this topic.

MECHANISMS AND DRIVE SYSTEMS: GEARING SYSTEMS

Gears are toothed wheels which are designed to transmit rotary motion and power from one part of a mechanism to another. Gearing systems are used to increase, or decrease, the output speed of a mechanism. They can also be used to change the direction of motion of the output. In this course, you will need to know about the following types of gear.

SIMPLE GEAR TRAIN

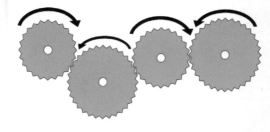

Gears work by interlocking or meshing the teeth of the gears together as shown. When two or more gears are meshed, they form a system known as a 'simple gear train'. The input gear which causes the system to move is called the driver. The output gear is known as the driven.

IDLER GEAR

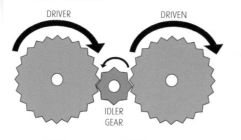

DRIVER   DRIVEN

IDLER GEAR

An idler gear is a gear that can be inserted into a system to allow the driver gear and the driven gear to rotate in the same direction. It has no effect on the speed of the driven gear.

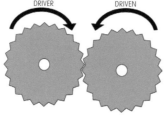

DRIVER   DRIVEN

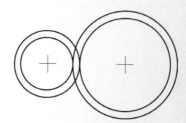

The drawing (above right) is the BSI Standard Convention symbol for a simple gear train. This symbol must be used when planning out gearing systems.

COMPOUND GEAR TRAINS

This drawing is the BSI Standard Convention symbol for a compound gear train. This symbol must be used when planning out gearing systems.

A compound gear system is multiple gear trains that are connected by a common axle. These are used when a very large change in speed is needed.

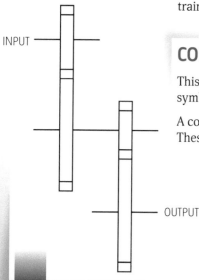

INPUT

OUTPUT

WORM AND WHEEL

Using a worm and wheel is a way of making a large speed reduction in a system as well as creating a large increase in torque. The worm (looks like a screw thread) is usually fixed to a driver shaft, or even directly to the motor shaft. It then meshes with a wheel that is fixed to the driven shaft. This means that the system will also run at 90° to the driver.

The worm wheel is seen as only having one tooth, which allows this huge reduction in speed, taking up very little space.

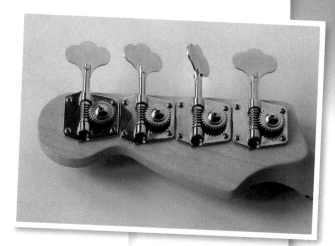

FRICTION WITHIN A SYSTEM

Within all these drive systems, a huge amount of friction can be created, causing heat. This can cause the material of the gears/pulleys to expand, which can cause a lot of damage to the drive system. To help reduce this, two things can be done:

1. Use a lubricant, such as oil or grease. By lubricating the moving parts, it allows the different parts to touch and move more easily without creating so much heat.

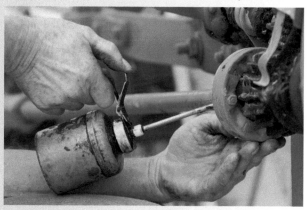

2. Use bearings. These are components that are designed to withstand wear. Ball and roller bearings are used within high-speed and high-force systems to replace any rubbing actions with rolling actions instead. Rolling doesn't create as much friction as rubbing does. These bearings can eventually be removed and replaced instead of damaging and replacing the whole system.

 ## THINGS TO DO AND THINK ABOUT

Create a word-search using keywords you have learned from this section. Give it to other people in your class to see if they can solve it.

 ONLINE

To learn more about gearing systems, visit www.brightredbooks.net

 DON'T FORGET

In the exam, you have to be specific about what parts will need to be lubricated. For example, stating 'lubricating the compound gears would reduce friction' would give you that mark, but just saying 'lubricate the system' will not.

DON'T FORGET

Knowing techniques to reduce friction will not only be useful for your exam, but also it is a huge part of this career path if you choose it.

 VIDEO LINK

Go to www.brightredbooks.net to watch a video on how roller bearings work.

 ONLINE TEST

Test your knowledge of this topic at www.brightredbooks.net

MECHANISMS AND DRIVE SYSTEMS: MULTIPLIER RATIOS

MOVEMENT MULTIPLIER RATIO

The ratio that can be created to show the change in speed of gears is known as the **movement multiplier ratio**. This can be found by dividing the number of teeth of the driven gear by the number of teeth of the driver.

$$MMR = \frac{Driven}{Driver}$$

This can then be used to calculate the output speed of the system.

EXAMPLE:

$MMR = \frac{Driven}{Driver} = \frac{12}{24} = \frac{1}{2}$ **or** 1:2

This means the driver will be going at half the speed of the driven. In other words, for every rotation the driver makes, the driven will make 2.

If the driver is rotating at 100 revs per minute, the driven will rotate at 200 rev min⁻¹ $\left(\frac{100}{\frac{1}{2}}\right)$.

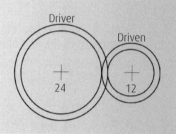

Gears can also be used to decrease the speed of a mechanism.

EXAMPLE:

$MMR = \frac{Driven}{Driver} = \frac{24}{12} = \frac{2}{1}$ **or** 2:1

This means the driver will be going at twice the speed of the driven. In other words, for every rotation the driver makes, the driven will only make half a rotation.

If the driver is rotating at 100 revs per minute, the driven will rotate at 50 rev min⁻¹ $\left(\frac{100}{\frac{2}{1}}\right)$.

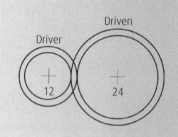

⚙ ACTIVITY 1:

Work out the movement multiplier ratio and speed of the following gears.

(a) Input speed of gearing system = 50 rev min⁻¹

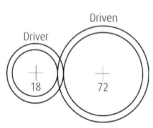

(b) Input speed of gearing system = 150 rev min⁻¹

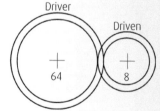

(c) Input speed of gearing system = 175 rev min⁻¹

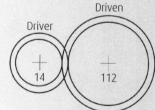

MOVEMENT MULTIPLIER RATIO IN COMPOUND GEARS

If a gearing system is required to produce a very large change in speed, problems could arise due to the size of gear wheels that would be needed if a simple gear train is used. This problem can be resolved by using compound gears.

To do this, we work out the movement multiplier ratio of each section of the gear system, then multiply them together.

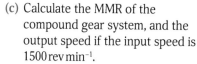
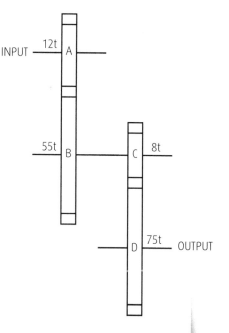

EXAMPLE:

If the input speed is 100 rev min⁻¹, calculate the output speed.

The movement multiplier ratio for the first pair of meshing teeth is:

Ratio of AB = $\frac{\text{driven}}{\text{driver}} = \frac{B}{A} = \frac{60}{20} = \frac{3}{1}$

The multiplier ratio for the second pair of meshing teeth is:

Ratio of CD = $\frac{\text{driven}}{\text{driver}} = \frac{D}{C} = \frac{80}{10} = \frac{8}{1}$

The total multiplier ratio is calculated by multiplying both ratios:

Total ratio = $\frac{3}{1} \times \frac{8}{1} = \frac{24}{1}$ = 24:1

Speed = $\frac{100}{24}$ = **4·2 rev min⁻¹**

 ACTIVITY 2:

(a) Calculate the MMR of the compound gear system, and the output speed if the input speed is 75 rev min⁻¹.

(b) Calculate the MMR of the compound gear system, and the output speed if the input speed is 2000 rev min⁻¹.

(c) Calculate the MMR of the compound gear system, and the output speed if the input speed is 1500 rev min⁻¹.

THINGS TO DO AND THINK ABOUT

a Using the correct symbols, draw out a compound gear system to achieve the lowest possible output speed. You can use the following gears: 46, 25, 15 and 8 teeth.

b If you achieve the output speed of 10 rev min⁻¹, what would your input speed be?

c Use Yenka to build and simulate this system.

ONLINE TEST

Test your knowledge of this topic at www.brightredbooks.net

MECHANISMS AND DRIVE SYSTEMS: SPEED AND TORQUE

VELOCITY RATIO

The overall ratio of a gearing system can also be worked out by only knowing the input and output speeds. This is known as the 'velocity ratio'. The velocity ratio for a gearing system is the ratio of the number of revolutions made by the input to the number of revolutions made by the output.

$$\text{velocity ratio} = \frac{\text{speed of input}}{\text{speed of output}}$$

EXAMPLE:

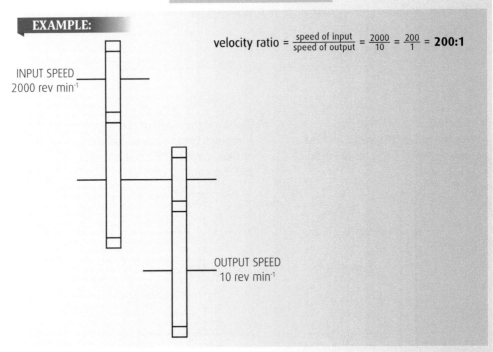

$$\text{velocity ratio} = \frac{\text{speed of input}}{\text{speed of output}} = \frac{2000}{10} = \frac{200}{1} = \mathbf{200{:}1}$$

INPUT SPEED
2000 rev min^{-1}

OUTPUT SPEED
10 rev min^{-1}

ACTIVITY 3:

Work out the velocity ratio for the following:

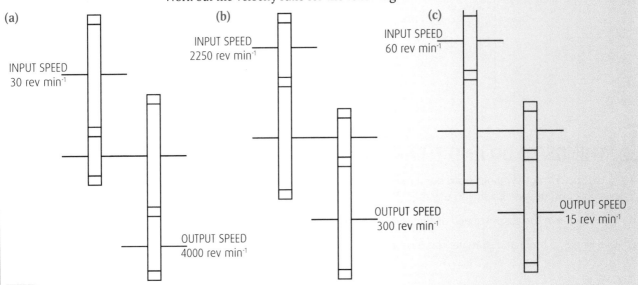

(a)

INPUT SPEED
30 rev min^{-1}

OUTPUT SPEED
4000 rev min^{-1}

(b)

INPUT SPEED
2250 rev min^{-1}

OUTPUT SPEED
300 rev min^{-1}

(c)

INPUT SPEED
60 rev min^{-1}

OUTPUT SPEED
15 rev min^{-1}

TORQUE

To turn an object in a rotary motion, such as turning or twisting, requires a certain amount of force. This force is known as 'torque'. To calculate torque, we use this equation:

Torque = Force × radius

EXAMPLE:

How much torque is required to tighten a nut if the force required is 55 N and the radius of the tool is 25 mm?

Torque = Force × radius

$\quad$ = 55 × 25

$\quad$ = 1375 Nmm

$\quad$ = **1.4 Nm⁻³ (2 S.F.)**

 **ACTIVITY 4:**

(a) A flag is raised by a small hand winch. The cord passes around a drum of 150 mm diameter at a force of 12·7 N. What is the torque needed?

(b) A drive system is used to open and close a sports stadium roof. In order to do this effectively, a force of 25 000 N is required from the final gear in the drive system. Calculate the required torque if the radius of the gear is 450 mm.

(c) A simple winding arrangement has been created on a building site to lift tools from one level to another. The winding drum has a diameter of 400 mm, and the total amount of tools being raised has a mass of 175 kg. What torque is needed to lift these?

 ## THINGS TO DO AND THINK ABOUT

Using appropriate symbols, draw out a compound gear system to achieve the lowest possible movement multiplier ratio. You can use the following gears: 44, 30, 15 and 10 teeth. Use Yenka to build and simulate this system.

 ONLINE TEST

Test your knowledge of this topic at www.brightredbooks.net

PNEUMATIC SYSTEMS: COMPONENTS 1

WHAT IS PNEUMATICS?

Pneumatics is something that you probably know very little about, yet it's more than likely that you come across it every day without even realising it. Pneumatics is similar to hydraulics, but instead of using fluid, it uses compressed air as its source of energy. Pneumatic systems are commonly used in industry, as they convert the stored energy within compressed air to provide movement in a range of components called cylinders.

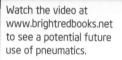

ONLINE

Watch the video at www.brightredbooks.net to see a potential future use of pneumatics.

Engineers commonly use pneumatics in industries such as medical, packaging, entertainment and robotics. Pneumatics can also be useful in very specific applications where there is a high chance of hazards – for example, in an area with gaseous chemicals. Electronics would be of no use because, if a stray spark of electricity arose, it could mean disaster and lives lost.

Pneumatics are being used increasingly, and in more imaginative ways that would have been unthinkable a decade or two ago. Creative applications, from robotics to pneumatic muscles, are consistently making the news. This shows not only the creativity of the engineering community, but also the inherent flexibility and adaptability of this important technology.

COMPONENTS

The equipment you will use in this area of engineering can be split up into two basic categories – cylinders and valves.

Cylinders are the 'muscles' of pneumatic systems, as they are used to move, hold or lift objects, or even to operate other pneumatic components. Cylinders are operated by compressed air, and they convert the stored energy in the compressed air into linear motion.

Valves are used to control the flow of compressed air to a cylinder. They can be used to turn the air on or off, or to change the direction in which the air is flowing, or even to slow down the airflow.

contd

3/2 valves

This component is known as a 3/2 valve – or, to be specific, a 'push-button actuated, spring-return 3/2 valve'. It gets its name because it has **3 ports** (3 connections) and **2 states** (on or off).

A port is where we can connect a pipe. In this example, our 3 ports are:

Port 1 – mains air   Port 2 – output connections   Port 3 – exhaust air

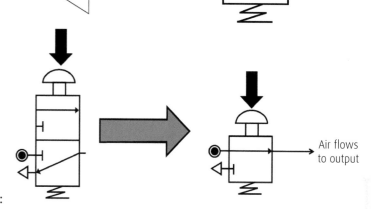

For this 3/2 valve to work, the push-button actuator is pressed. When this is pressed, imagine that the top square gets pushed down. This will then replace the bottom section and connect the mains air to the output, allowing air to flow. When the button is released, the spring-return actuator will push it back up, reconnecting the output to the exhaust, stopping air from flowing.

Air flows to output

Actuators

Actuators are what will switch the valve on or off. There are a number of different ways we can operate a 3/2 valve:

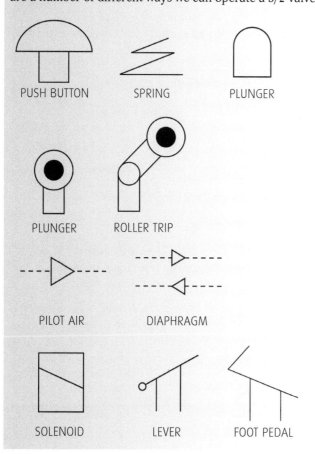

PUSH BUTTON   SPRING   PLUNGER

PLUNGER   ROLLER TRIP

PILOT AIR   DIAPHRAGM

SOLENOID   LEVER   FOOT PEDAL

DON'T FORGET

When naming a component, you have to state its **full** name. That means stating what actuators are used: for example, a 'roller-trip actuated, spring-return 3/2 valve'. Look at the top and bottom of the valve to see what actuators are used.

THINGS TO DO AND THINK ABOUT

Create a memory game, like the card game 'Pairs', except that one card is a picture, and the other card is the name of the symbol. Play this with classmates to improve your recognition of the symbols used so far. As you progress through this unit, you can add new cards.

ONLINE TEST

Test your knowledge of the components of pneumatics systems at www.brightredbooks.net

PNEUMATIC SYSTEMS: COMPONENTS 2

SINGLE-ACTING CYLINDER

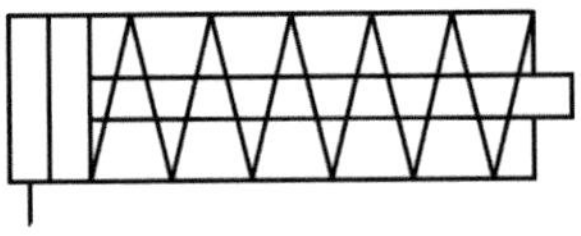

A single-acting cylinder requires only one air supply for it to work. When air is put into the cylinder, it will 'outstroke'. If the air stops, it will 'instroke'. It will do this automatically, due to the spring being in place forcing it back.

This can then be connected to other components to create movement.

EXAMPLE:

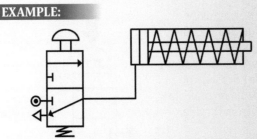

As the push-button, spring-return 3/2 valve is pressed, the single-acting, spring-return cylinder will outstroke. Once you stop pressing it, the cylinder will instroke.

A circuit like this can also be expanded to create a pneumatic version of an AND gate. By adding another push-button 3/2 valve, the first **and** second valves need to be pressed for the single-acting cylinder to outstroke.

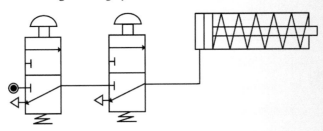

ONLINE

Follow the link at www.brightredbooks.net to simulate some pneumatic circuits.

SHUTTLE VALVE

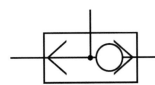

OR gate control is also possible with 3/2 valves and a single-acting cylinder. This is useful, as sometimes we need to control a pneumatic circuit from more than one position. To do this, we need to use another component called a shuttle valve.

A shuttle valve is used to change the direction of air in a circuit. A small ball inside the component gets blown from side to side, allowing air to pass through to the cylinder.

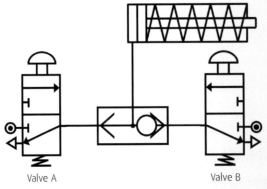

Valve A   Valve B

If valve A **or** valve B is pressed, it will cause the single-acting cylinder to outstroke. Once the buttons are released, it will instroke.

DOUBLE-ACTING CYLINDER

Unlike a single-acting cylinder, this **does not** have a spring inside to return it to its original position. Instead, it has 2 air supplies – one that will cause it to outstroke, and another that will cause it to instroke.

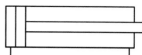

As a push-button, spring-return 3/2 valve (valve A) is pressed, the double-acting cylinder outstrokes. Once the other push-button, spring-return 3/2 valve (valve B) is pressed, the cylinder will instroke.

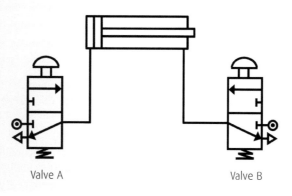

Valve A   Valve B

5/2 VALVE

Controlling a double-acting cylinder in this way with two 3/2 valves can bring up many problems. The main one is that, after you have actuated the 3/2 valve, it returns to the off state. This means the air stops being supplied to the cylinder, so the pressure stops, and the force is no longer there. This will mean the cylinder can be pushed back/pulled out by hand.

Another disadvantage is that the 3/2 valve would need to be continually actuated until the cylinder is fully out/instroked. If the valve is released before this, the piston will stop short of its final position. To solve these problems, a 5/2 valve can be used.

This gives greater control over a double-acting cylinder. This works in the same way as a 3/2 valve, except there are 2 exhaust ports and 2 output connections.

OUTPUT
CONNECTIONS

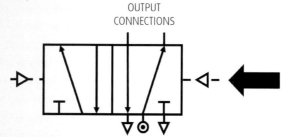

If air is supplied from this side, it will connect using that side's connections, sending output air to the right-hand-side output connection, and allowing air to come in from the other output, going straight through to the exhaust.

OUTPUT   OUTPUT
CONNECTIONS   CONNECTIONS

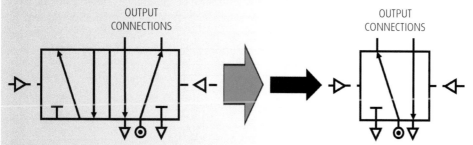

If the air is supplied from the other side, that side of the valve 'slides' over and creates new connections to the output, mains air and exhausts. This means that air will now be supplied to the left-hand output connection, and air can come back into the valve through the right connection, and into the exhaust.

DON'T FORGET

A 5/2 valve is usually actuated by pilot air, as shown by a dashed line. Pilot air is a short burst of air that will activate the valve.

ONLINE

Visit www.brightredbooks.net to get a greater understanding of pneumatic components.

THINGS TO DO AND THINK ABOUT

Within a factory, pneumatic valves are used to stamp aluminium into different shapes. As a safety feature, the circuit must be designed so that both the technician's hands are clear of the valves. Design a circuit that could do this.

ONLINE TEST

Test your knowledge of the components of pneumatics systems at www.brightredbooks.net

PNEUMATIC SYSTEMS: COMPONENTS 3

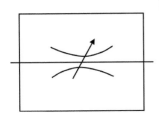

SPEED CONTROL

So far, every circuit we have looked at will cause the pistons to move very quickly. In a real-world situation, this could be very dangerous or could even cause the circuit to stop working correctly. To solve this problem, we can use certain components to control the flow of air through the valves.

One way of doing this is using a **restrictor**.

This will slow down the air in both directions, meaning it will slow down the instroke and the outstroke. This is essentially like a screw going into the wire. If the screw is tightened, it gives a smaller area for the air to get through, hence slowing it down.

The other way of controlling speed is by using a **unidirectional restrictor**.

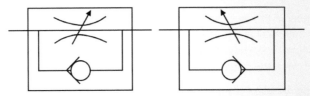

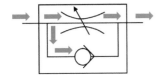

This will slow down the air in 1 direction only. This is dependent on the bottom section. As the air flows through the component, it gets split between 2 paths. If the air flows from this direction, the ball is blown into the valve, blocking this path for the air to follow. This in turn will slow it down.

If it flows from the other direction, the ball is instead blown away from the entrance, allowing air to flow unrestricted.

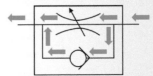

EXAMPLE:

The unidirectional restrictor is placed so that it will slow down the exhaust air coming from the cylinder. When valve A is pressed, the 5/2 valve will change state and start to supply the double-acting cylinder with air, causing it to outstroke. Air trapped on the other side of the cylinder will escape slowly because of the unidirectional restrictor. This in turn causes the piston to outstroke slowly and smoothly without affecting the force exerted.

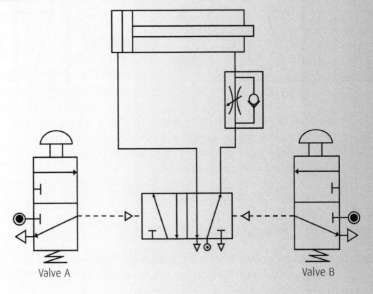

Valve A   Valve B

TIME DELAY

Sometimes it is desired for there to be a time delay between when the valve is actuated and when the cylinder responds. This pause can be created by a component known as a reservoir, and it is frequently used in conjunction with a unidirectional restrictor.

This is simply a container for the compressed air. By connecting this to a pipe, it increases the space that has to be pressurised before the next component is operated. This will create a time delay.

ONLINE

Go to www.brightredbooks.net to improve your knowledge of time-delay circuits.

AUTOMATIC CIRCUITS

Automatic circuits are how pneumatics are mostly used in industry. They not only help to speed up production, but also they allow for uniformity, making sure that goods are all made to the same standard. There are two types of automatic circuit that could be used:

Semi-automatic circuits

A semi-automatic circuit is one which will complete a set process once a human operator has started it.

Within this circuit, the process will start when the operator presses the push-button on the 3/2 valve (valve A). This will change the state of the 5/2 valve, causing the double-acting cylinder to outstroke. The cylinder, when outstroked far enough, will activate the roller-trip-actuated 3/2 valve (valve B). This will then change the state of the 5/2 valve, causing the cylinder to instroke. The process is then ready to begin again.

Valve A

Valve B

Fully automatic circuits

A fully automatic circuit is one that will continue to work, performing the task over and over again, without manual intervention. When mains air is supplied, it will start and will continue to work.

This circuit works by the double-acting cylinder instroking. This will activate the roller on the 3/2 valve (valve A) and cause the 5/2 valve to change state. This will then outstroke the double-acting cylinder. When it is fully outstroked, it trips the roller on the other 3/2 valve (valve B). This will change the state of the 5/2 valve again, causing the cylinder to instroke. The process then begins over, and will continue to operate in this fashion.

A fully automatic circuit can be interrupted, though, in case of an accident or emergency. This can be done by putting a lever-lever 3/2 valve in one pilot line. This can then act as an on/off switch.

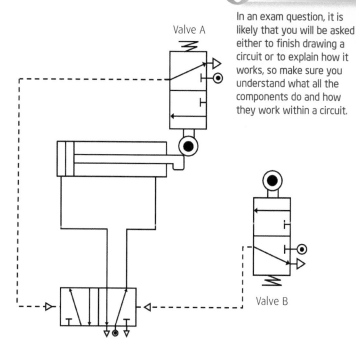

Valve A

Valve B

DON'T FORGET

In an exam question, it is likely that you will be asked either to finish drawing a circuit or to explain how it works, so make sure you understand what all the components do and how they work within a circuit.

THINGS TO DO AND THINK ABOUT

A stamping machine is required to put stickers onto packaging. Within this system, it must have a slight delay, then outstroke fast. It must then instroke slowly. Design a fully automatic circuit that will do this; and then describe, using the proper terminology, how the circuit works.

ONLINE TEST

Test your knowledge on this section at www.brightredbooks.net

PNEUMATIC SYSTEMS: PNEUMATIC CIRCUITS

DESCRIBING A CIRCUIT

In an exam, it is very possible you will be asked to describe how a circuit works, and it will be worth quite a lot of marks. Although these can be daunting, they are nothing to fear if you just start at the beginning and follow the logical path.

EXAMPLE:

A pneumatic system is used to control models in a living museum. Part of the circuit is shown below.

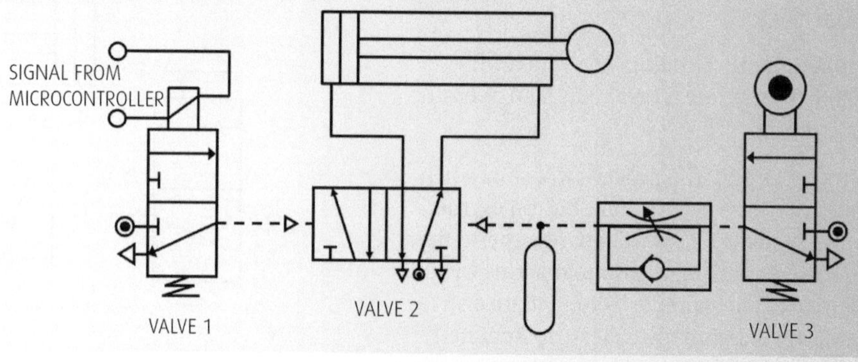

CYLINDER A

SIGNAL FROM MICROCONTROLLER

VALVE 1

VALVE 2

VALVE 3

Describe, using the appropriate terminology, the operation of the circuit.

To start answering this question, you first have to find the starting point of the circuit. Look for an actuator that has to be pushed or controlled manually – on this circuit, it will be 3/2 Valve 1. You then have to follow the logical path and think what happens when this is pushed, and so on.

One approach to a question like this is writing what happens at each step as a bullet-point list. This will break it up, and you won't get confused writing the description – or, if you do, you can easily find your place again.

- When a signal is sent from the microcontroller, the solenoid, spring-return 3/2 Valve 1 is actuated.
- This sends pilot air to the pilot-pilot 5/2 Valve 2.
- This will cause Cylinder A to outstroke.
- This will hit the roller on the roller, spring-return 3/2 Valve 3.
- After a delay, this will send pilot air back to 5/2 Valve 2.
- This causes Cylinder A to instroke.
- The system is now reset and waiting for another signal from the microcontroller.

EXAMPLE:

A company uses cylinders to press metal dies into hot plastic to create parts for a children's toy. As each cylinder is lowered individually, it presses the plastic into shaped recesses.

The sequence of operations for this process is as follows.

- An operator pushes a button to start the process
- Cylinder A lowers
- Cylinder B lowers
- Cylinder A raises
- Cylinder B raises.

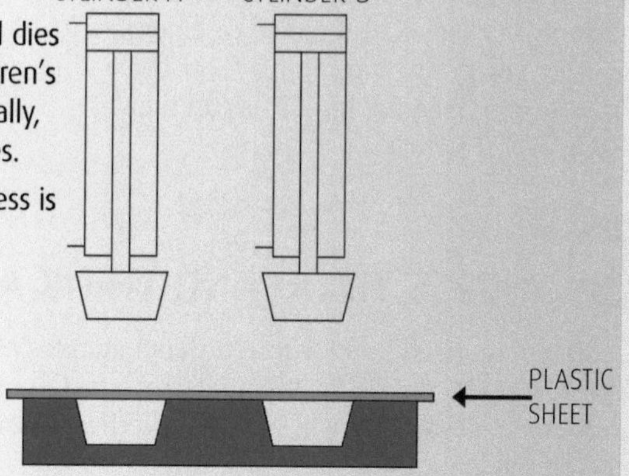

CYLINDER A   CYLINDER B

PLASTIC SHEET

DON'T FORGET

When naming components, make sure you mention their full name as well as the name given (e.g. Valve A). If you have to refer to that valve again, you then only have to call it the given name, and not the full one.

DON'T FORGET

Sometimes it will describe a cylinder as + or -. If it is +, this means it is outstroking. If it is -, it means it is instroking.

contd

The diagram for the circuit is shown below. Using the proper terminology, describe how the circuit works.

As before, you first have to find the starting point of the circuit. Look for an actuator that has to be pushed or controlled manually – on this circuit, it will be 3/2 Valve A. You then have to follow the logical path and think what happens when this is pushed, and so on.

The way this circuit works is as follows:
- When the push-button, spring-return 3/2 Valve A is activated, it will change the state of the pilot-pilot 5/2 Valve B.
- This will cause the double-acting cylinder (Cylinder 1) to outstroke.
- This will hit the roller on the roller-actuated, spring-return 3/2 Valve C.
- This will then send pilot air to the pilot-pilot-actuated 5/2 Valve D.
- This causes the valve to change state, sending air to the double-acting cylinder (Cylinder 2), causing this to outstroke.
- This will hit the roller on the roller-actuated, spring-return 3/2 Valve E, sending pilot air to 5/2 Valve B.
- This causes Cylinder 1 to instroke, hitting the roller on 3/2 Valve F.
- This in turn sends pilot air to 5/2 Valve D, instroking Cylinder 2.
- The system has now reset and waits for a user to press the button again.

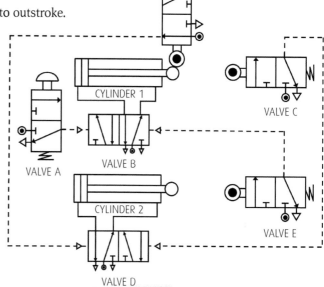

THINGS TO DO AND THINK ABOUT

Describe how this pneumatic circuit works, using the proper terminology.

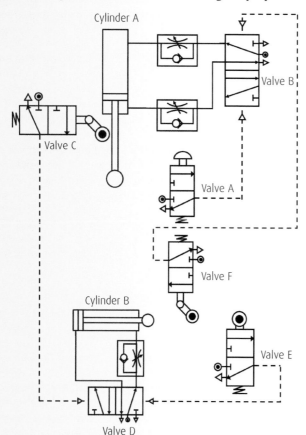

ONLINE TEST

Test your knowledge on this section at www.brightredbooks.net

PNEUMATIC SYSTEMS: OUTPUT FORCES FROM A CYLINDER

FORCE IN A SINGLE-ACTING CYLINDER

DON'T FORGET

The unit can change to suit the question. For example, if the question uses millimetres, your answer may be in Nmm^{-2}.

As we know, pneumatic components are circuits that are controlled by pressured air. This air pressure is measured in Nm^{-2}. We know the pressure going through the system, as there will be a pressure gauge on the compressor that supplies the air. These can be found on any system that relies on compressed gases or fluids. For example, it is more than likely there will be one on the boiler in your house for the central heating. This helps to detect leaks, as the pressure in the system would begin to fall if air was escaping from the pipes.

The force you get out of a cylinder depends on the air pressure inside it, and on the size of the piston. To calculate this pressure, we multiply the pressure by the area of the piston. This is measured in Newtons.

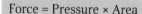

Force = Pressure × Area

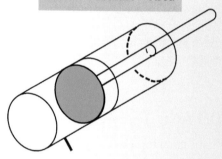

As you can see, the surface of the cylinder is circular. This means we have to use one of these calculations to work out the area of the piston:

$$\text{Area} = \pi r^2 \quad \text{or} \quad \frac{\pi d^2}{4}$$

FORCE IN A DOUBLE-ACTING CYLINDER

As we know, a double-acting cylinder is of more use in practical applications due to the fact that both the input and output can be controlled by compressed air. However, the outstroke force is greater than the input force. This is because, when it is instroking, the piston is at this side, taking up space. This means that there is less area for the compressed air to fill, meaning less force created.

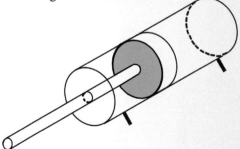

Therefore, we have to find the 'effective area'. This is done by calculating the area of the piston rod and subtracting it from the area of the piston.

Effective area = piston area – piston-rod area

contd

EXAMPLE:

A car-park barrier uses a double-acting cylinder to raise and lower the barrier. The cylinder has a diameter of 60 mm, with the piston rod being 20 mm in diameter. The air pressure is 0·7 N/mm². What forces are produced when the piston outstrokes and instrokes?

Outstroke

Area $= \frac{\pi d^2}{4} = \frac{3 \cdot 14 \times (60 \times 60)}{4} = \frac{3 \cdot 14 \times 3600}{4} = \frac{11\,304}{4} = 2826\,mm^2$

Force = Pressure × Area = 0·7 × 2826 = **1978·2 N**

Instroke

Cylinder area = 2826 mm²

Cylinder-rod area $= \frac{\pi d^2}{4} = \frac{3 \cdot 14 \times (20 \times 20)}{4} = \frac{3 \cdot 14 \times 400}{4} = 314\,mm^2$

Effective area = Piston area – Piston-rod area = 2826 – 314 = 2512 mm²

Force = Pressure × Effective Area = 0·7 × 2512 = 1758·4 N

 **ACTIVITY:**

(a) A piston has a diameter of 30 mm and is supplied with air at a pressure of 3·2 Nmm⁻².

   Calculate the outstroking force of the cylinder.

(b) The cylinder shown is used to open and close a greenhouse window.

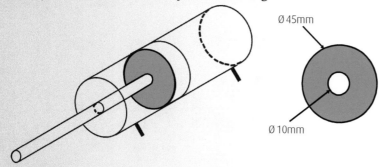

Ø 45mm

Ø 10mm

Calculate:

   (i) the effective area of the piston when it instrokes.

   (ii) the force supplied to the cylinder when the air pressure is 0·7 Nmm⁻².

(c) A pneumatic cylinder is shown.

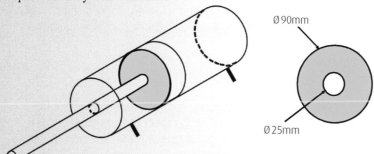

Ø 90mm

Ø 25mm

Calculate the air pressure required to produce an instroking force of 2450 N.

 ONLINE

Visit http://www.bbc.co.uk/schools/gcsebitesize/design/systemscontrol/pneumaticsrev6.shtml to develop your knowledge of pressure within a cylinder.

 THINGS TO DO AND THINK ABOUT

Go to the SQA past-papers website (http://www.sqa.org.uk/pastpapers/findpastpaper.htm?subject=Engineering+Science&level=N5) and try to answer some of the pneumatics questions. The answer schemes are also available on this site so that you can mark yourself and see what you may be doing wrong.

 ONLINE TEST

Test yourself on output forces at www.brightredbooks.net

STRUCTURES: STRENGTH OF STRUCTURES

ONLINE

Follow the link at
www.brightredbooks.net
to get a greater
understanding of a career in
structural engineering.

WHAT IS STRUCTURAL ENGINEERING?

Structural engineers help to create the constructed world that's all around us – whether this is bridges, tunnels or buildings.

Structural engineers are concerned with all aspects of a structure and its stability, meaning that they must be specialists in design, construction, repair, conversion and conservation. They are an integral part of a design and construction team, working alongside civil engineers, architects and other professionals to create a safe working structure.

Every structure has to deal with the conditions in which it is built. Houses in Norway and Canada will have to deal with large amounts of snow. Their roads have to deal with being frozen for prolonged periods of time. Buildings near the edge of tectonic plates have to deal with earthquakes. Bridges all around the world will need to carry different kinds of weights and loads. It is a structural engineer's job to consider all possible factors.

WHAT IS A STRUCTURE?

A structure can be defined as something that has an arrangement of parts that are joined together in a manner which provides some sort of strength. This means it is able to carry some form of load. There are many different types of structure from buildings to bridges, from chairs to vehicles.

The factors which contribute to the strength in a structure are:

- the materials used
- the shapes of the parts
- the methods used to join the parts together
- the manner in which they are arranged.

MATERIAL CHOICE

One thing a structural engineer would have to consider is the material being used – and this decision could include several different engineers to ensure that the correct material has been chosen. For example, a chemical engineer may be part of this discussion, as they would have the knowledge of how the material would react in different conditions, or what finish could be used to help protect it from the elements.

If you are asked to choose a material or a potential material for an engineering challenge, there are several things you could base this decision on.

First, how much will it cost? Engineers are given a budget when designing an engineering solution, and this has to be stuck to.

Second, will it corrode? For example, if you decided that steel was the material you would use, this will rust. Does this mean it will need to be painted or given some sort of rustproof finish?

Third, what are the properties of the material, and will it be suitable? Every material has strengths and weaknesses that will affect its performance.

Strength

A strong material can resist force. Some materials, like mild steel, are strong in tension but weak in compression. The opposite is also true for some materials that are strong in compression but weak in tension, like concrete. This is one of the reasons that concrete is often reinforced with mild steel.

Elasticity

An elastic material can return to its original length or shape once a load or force has been removed. For example, rubber is seen as having high elasticity.

Plasticity

A material with plasticity can change its shape or length under a load and can stay deformed even when the load is removed. 'Play-doh' has high plasticity.

Brittleness

A material is brittle if it can be easily cracked, snapped or broken.

Toughness

A tough material can absorb a sudden sharp load without causing permanent deformation or failure. Tough materials also need to have high elasticity.

Hardness

A hard material can resist erosion or surface wear.

Ductility

A ductile material can be stretched without breaking, and can be formed into shapes such as very thin sheets or very thin wire. Copper is seen as ductile.

You do not need to remember these properties in National 5, but they are good reference if you have to decide on a material for an engineering solution.

DON'T FORGET

If you are ever asked to choose a material for a potential engineering solution, always make sure you have analysed the requirements and compared a minimum of 2 suitable possibilities – and then make a choice, justifying that decision.

THINGS TO DO AND THINK ABOUT

Copy and complete this table.

Add extra materials, and think of their uses and what the material properties would be.

Material	Uses	Material properties
Nylon fibres	tights	elastic
Polyethylene	washing-up bowl	tough, brittle
Rubber		
Wood		

ONLINE TEST

Head to www.brightredbooks.net to test yourself on this topic.

STRUCTURES: LOADS AND FORCES

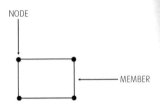

NODE

MEMBER

FRAME STRUCTURES

A frame structure is the most common type used, and this type of structure can be broken up into 2 parts: the 'members', which are the beams that make up the structure, and the 'nodes', which join them together.

Depending on where a load is placed, the structure will act differently.

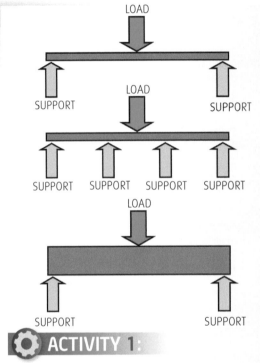

LOAD

SUPPORT   SUPPORT

LOAD

SUPPORT   SUPPORT   SUPPORT   SUPPORT

LOAD

SUPPORT   SUPPORT

LOADS ON A MEMBER

Putting weight on a structure can cause it to change shape and eventually to fail. Putting weight on a member could cause it to bend.

This can be prevented by adding extra supports.

It can also be prevented by changing the thickness of the member.

These are both valid ways of strengthening a member; but, as an engineer, you also have to consider possible consequences. Doing this will add extra weight, and although it would solve our original problem, it could lead to many other follow-on problems.

⚙ ACTIVITY 1:

By using a standard 300 mm ruler and some pencils, test how loads act on a member. Have the ruler lying down flat like in the first two examples, and put weight on it. Reposition it now so that the ruler is on its edge, and put weight on it. Does it act in the way stated and expected? Write down the results ... what happened, and why?

LOADS ON A NODE

Due to the difficulties that can occur when a load is placed on a member, putting weight on a node should be considered instead. This is the strongest place on a structure to put a load. By doing this, though, it can distort the shape.

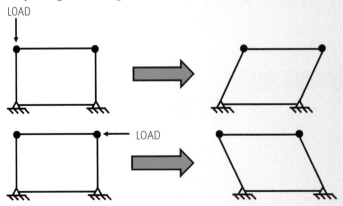

LOAD

LOAD

contd

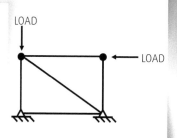

LOAD

LOAD

⚙ ACTIVITY 2:

Prove this by building this model with Lego, Fischertechnik or lollipop sticks and split pins. Holding the bottom two corners, get someone else to press down on the structure from different directions. Does the structure move?

This problem can be resolved using a technique known as **triangulation**. By inserting a diagonal member into the structure and changing the shape from a rectangle into two triangles, it will strengthen the structure. Triangles are the strongest shape and should be utilised when designing any structure.

⚙ ACTIVITY 3:

Using the same structure you have built previously, prove that triangulation works by adding an extra beam. Does the stability and strength of the structure improve?

FORCES ON MEMBERS

By loading on a node, it can cause a member to react in different ways. These may not always be visible, but they will happen.

If the ends of a member are pushed together, it will become shorter and the middle will become fatter.

Push ➡  ⬅ Push

This type of load is known as **compression**. The member is under a force known as **stress**, and the member is known as a **strut**.

If the ends of a member are being pulled apart, it will become longer and the middle will become thinner.

Pull ⬅  ➡ Pull

This type of load is known as **tension**. The member is under a force known as **strain**, and the member is known as a **tie**.

> **EXAMPLE:**
>
> Within this example, **member A** will be a **tie**, as the load will cause it to be pulled, and under **tension**.
>
> **Member B** will be a **strut**, as the load is pulling down, causing it to get squashed, or be put under **compression**.

Member A

Member B

ONLINE

Follow the link at www.brightredbooks.net to download and play West Point Bridge Designer. Here you can strengthen your knowledge of structures by designing bridges, as well as entering a virtual bridge-building competition.

DON'T FORGET

In the exam paper, it may ask you to name the members of a frame. If you are unsure, draw a diagram showing where the structure is being held, and where the force is pulling or pushing. This makes it easier to work out what is happening in each individual member.

⚠ THINGS TO DO AND THINK ABOUT

Research a local bridge, and research what types of engineers would have been involved in its construction. Draw a free-body diagram of the structure, then build a model of the structure to show how triangulation has been used to strengthen it or to improve its stability or rigidity.

ONLINE TEST

Test your knowledge of this topic online at www.brightredbooks.net

STRUCTURES: STRESS AND STRAIN

STRESS

When building structures, an engineer has to take **stress** into consideration. Stress is when a force is put onto a member to cause **compression**. This is important for engineers to know because, if too large a load is placed on a structure, it will bend, buckle or break.

The stress can vary either by the force applied, or by the size of the structure. For example, if the same load is applied to column A and column B, then column A will suffer more from stress. This is because column A has a far smaller cross-sectional area.

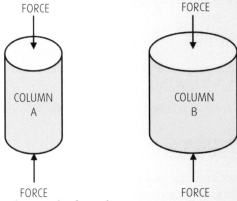

Stress can be calculated using the formula:

$$\text{Stress} = \frac{\text{Force}}{\text{Area}} \quad \sigma = \frac{F}{A}$$

Force is measured in Newtons. Area is measured in m². Therefore, stress is measured in Nm⁻².

EXAMPLE:

An oil rig is designed so that each of its four legs always carries an equal share of its 160 MN load. The legs are 20 m in diameter. Work out the stress on each leg.

Total force = 160 MN

Each leg = $\frac{160}{4}$ = 40 MN

$\sigma = \frac{F}{A} = \frac{40}{(3\cdot14 \times 20^2)/4} = \frac{40}{(3\cdot14 \times 400)/4} = \frac{40}{(12560/4)} = \frac{40}{314} = 0\cdot127\,\text{Mn}^{-2}$

Stress = 127 kN²

contd

ACTIVITY 1:

(a) A 25 × 25 mm square bar is subjected to a tensile load of 750 N. Calculate the stress in the bar.

(b) A shop sign with a force of 370 N is held up by 4·5 mm wire. What is the stress on this wire?

(c) A 10 kN load is held up by steel wire. The stress on this wire should not exceed 275 N/mm². Calculate the minimum diameter of wire required to support the load.

STRAIN

When building structures, an engineer also has to take **strain** into consideration. Every material will change shape to some extent when a force is applied to it. If a compressive load is applied, then the length will reduce. If a tensile load is applied, then the length will increase.

Strain is when a load causes a member to elongate in size. This can vary depending on the length of the material, and by how much it is being deformed.

VIDEO LINK

Watch the video at www.brightredbooks.net to get a greater understanding of how stress and strain work.

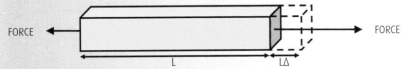

FORCE ← → FORCE

L   LΔ

Strain can be calculated using the formula:

$$\text{Strain} = \frac{\text{Change in length}}{\text{Original length}} \qquad \varepsilon = \frac{\Delta L}{L}$$

Strain does not have a unit.

EXAMPLE:

A boat is tied to a mooring post by polypropylene ropes. The rope is 15 m long, and under force it is extended by 32 mm. Calculate the strain the rope is under.

$$\varepsilon = \frac{\Delta L}{L} = \frac{32}{15\,000} = \mathbf{0\cdot002}$$

 ACTIVITY 2:

(a) A load is lifted using a wire that is 7 m long. When the load is applied, the wire stretches by 3·7 mm. Calculate the strain on the wire.

(b) A steel bar 4 m long has an allowable strain of 0·0063. Calculate the change in length.

(c) During testing, it was found that a steel rod stretched 0·75 mm, with a strain of 0·0022. What is the original length of the rod?

THINGS TO DO AND THINK ABOUT

Why is it important for an engineer to know how a material behaves under stress and strain? And why do engineers test materials to find out how they behave under these conditions? Research this on the internet, and make notes explaining why.

ONLINE TEST

Head to www.brightredbooks.net to test yourself on this topic.

STRUCTURES: FREE-BODY DIAGRAMS AND MOVEMENTS

FREE-BODY DIAGRAMS

Within a structure, an engineer has to know all forces acting on it to ensure that it does not collapse. This is done by drawing a **free-body diagram**. This is a representation of the structure drawn only with lines, arrows, dimensions and forces.

EXAMPLE:

By removing all the visual components and replacing them with their force values, it allows a better understanding of how the forces can affect the structure. This means the engineer would be able to ensure that the structure can support the loads that are affecting it.

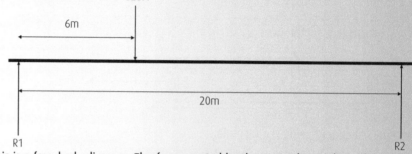

This is a free-body diagram. The force created by the person's weight is shown (in Newtons), as well as the distance from the end of the structure. The total length of the structure should also be shown, as well as any supports that exist, shown as reaction points R1 and R2.

> **DON'T FORGET** ➕
>
> All known information must be shown on a free-body diagram. If it gives you a weight in kg, you must change that to Newtons by multiplying it by the force of gravity (9·8).

⚙️ ACTIVITY 1:

(a) As part of an army training routine, soldiers have to climb across monkey bars 25 metres long. A soldier weighing 95 kg stops 9 metres along to have a rest. Draw the free-body diagram.

(b) Draw the free-body diagram.

(c) Draw the free-body diagram.

MOMENTS

The turning effect that can occur on a structure due to force is called a **moment**. Each moment is measured in **Newton metres**.

This has to be calculated to ensure that the structure is in equilibrium. When any system is in perfect balance and is in a steady state, it is said to be in equilibrium.

The load in this system wants to turn the beam clockwise, so the effort must try to turn the system anti-clockwise. To ensure that the beam is in equilibrium, these moments must be equal.

contd

A moment is calculated by multiplying the force exerted on it by the distance it is from the pivoting point, or fulcrum.

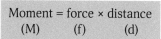

Moment = force × distance
(M)   (f)   (d)

So, this means that we can calculate the moments of the structure by ensuring that the total clockwise moments equal the total anti-clockwise moments.

↻ ↺   Σ   M = Σ   M

$\Sigma (f \times d) = \Sigma (f \times d)$

EXAMPLE:

Moments with 1 load at each side.

Calculate the moments to work out what x is.

↻ ↺   Σ   M = Σ   M

$(f \times d) = (f \times d)$

$600\,N \times 3\,m = x\,N \times 2\,m$

$1800\,N = 2x$

$x = \frac{1800}{2} = 900\,N$

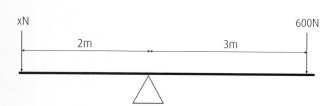

EXAMPLE:

Moments with multiple loads at each side.

Calculate the moments to work out x.

↻ ↺   Σ   M = Σ   M

$\Sigma (f \times d) = \Sigma (f \times d)$

$(300\,N \times 4\,m) + (200\,N \times 6\,m) = (100\,N \times 1\,m) + (x\,N \times 4\,m)$

$1200 + 1200 = 100 + 4x$

$4x = 2300$

$x = \frac{2300}{4} = 575\,N$

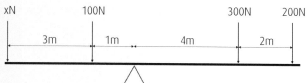

VIDEO LINK

Go to www.brightredbooks.net to help your understanding of moments.

DON'T FORGET

When calculating moments, the distance the force is hitting **always** comes from the pivot point.

VIDEO LINK

Watch the clip at www.brightredbooks.net to develop your knowledge of moments on a beam.

ACTIVITY 2:

Calculate the missing values in the diagrams.

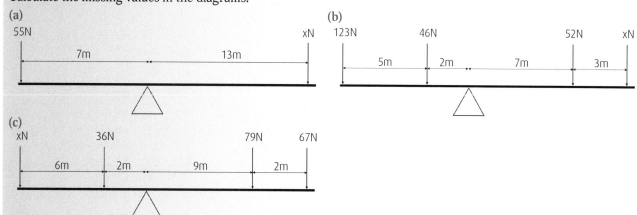

(a)

(b)

(c)

THINGS TO DO AND THINK ABOUT

Look at several structures around your home and local area. Try to draw a free-body diagram for each of them, including all relevant forces and dimensions. If there are unknown quantities, make a reasoned estimate as to what they might be.

ONLINE TEST

Test your knowledge of this topic at www.brightredbooks.net

STRUCTURES: MOMENTS AND REACTIONS

MOMENTS AND REACTIONS ON A STRUCTURE

By using the information on a free-body diagram, the forces at R1 and R2 can be calculated using something called **moments and reactions**. These need to be calculated to ensure that R1 and R2 are equal and opposite to the forces that are acting upon them. This is to ensure equilibrium – and, as we know, if equilibrium is not achieved in a structure, it will collapse.

As you have previously learned, moments are the turning forces within an object – and, for an object to be in equilibrium, these turning forces must balance each other out.

DON'T FORGET

Be careful of the wording in the question. If it says 'take moments around R1', this means R1 will be your pivot point. If it says 'take moments around R2', then R2 will become your pivot point.

$$\circlearrowright \quad \circlearrowleft \quad \Sigma \quad M = \Sigma \quad M$$
$$\Sigma (f \times d) = \Sigma (f \times d)$$

or $\quad \Sigma M = 0$

EXAMPLE:

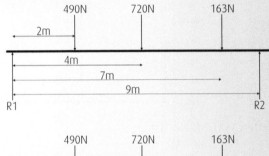
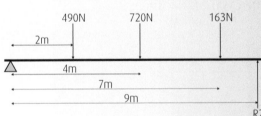
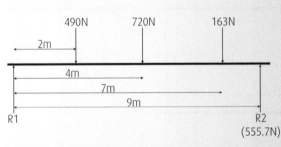

It is likely you will be given a question like this, where you have to calculate the forces at R1 and R2.

This will likely be worded along the lines of 'by taking moments about R1, calculate the size of the reaction R2'. This would mean that R1 would act as the pivoting point in the structure. This means we could now calculate what R2 is, using moments.

$$\circlearrowright \quad \circlearrowleft \quad \Sigma \quad M = \Sigma \quad M$$
$$\Sigma (f \times d) = \Sigma (f \times d)$$

$(490\,N \times 2\,m) + (720\,N \times 4\,m) + (163\,N \times 7\,m) = (R2 \times 9\,m)$

$980 + 2880 + 1141 = 9R2$

$5001 = 9R2$

$R2 = \frac{5001}{9}$

$R2 = \mathbf{555 \cdot 7\,N}$

Now we know the value of R2, we can calculate R1 by considering the vertical forces. Like moments, these have to be in equilibrium.

$$\uparrow \quad \downarrow \quad \Sigma = \Sigma$$

$R1 + R2 = 490\,N + 720\,N + 163\,N$

$R1 + 555 \cdot 7 = 1373\,N$

$R1 = 817 \cdot 3\,N$

Therefore the reactions of this supporting beam are:

$R1 = 555 \cdot 7\,N$ and $R2 = 817 \cdot 3\,N$

contd

EXAMPLE:

(i) By taking moments about R2, calculate the size of the reaction at R1.

(ii) Determine the size of the reaction at R2.

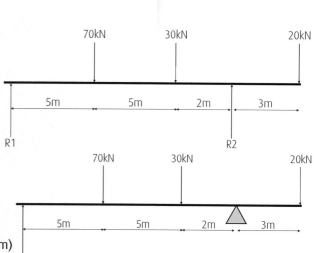

(i)

$\circlearrowleft \quad \circlearrowright \quad \Sigma \quad M = \Sigma \quad M$

$\Sigma (f \times d) = \Sigma (f \times d)$

$(R1 \times 12\,m) + (20\,kN \times 3\,m) = (30\,kN \times 2\,m) + (70\,kN \times 7\,m)$

$12R1 + 60 = 60 + 490$

$12R1 = 550 - 60$

$12R1 = 490$

$R1 = \frac{490}{12} = $ **40·8 kN**

$\uparrow \quad \downarrow \quad \Sigma = \Sigma$

$R1 + R2 = 70\,kN + 30\,kN + 20\,kN$

$40·8\,kN + R2 = 120\,N$

$R2 = 120\,kN - 40·8\,kN$

R2 = 79·2 kN

Therefore the reactions of this supporting beam are:

R1 = 40·8 kN and R2 = 79·2 kN

ACTIVITY:

In each of the systems below:

(i) Draw the free-body diagram.

(ii) By taking moments about R1, calculate the size of the reaction at R2.

(iii) Determine the size of the reaction at R1.

ONLINE

Visit www.brightredbooks.net to develop your knowledge of moments within a structure.

(a)

(b)

(c)
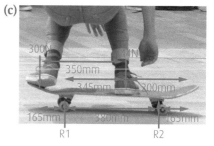

THINGS TO DO AND THINK ABOUT

Draw your own free-body diagrams, making up the numbers, and determine the sizes of R1 and R2. Give these free-body diagrams to some of your classmates. Can they work out the solutions? If your answers differ, discuss these and work out why and what has gone wrong.

ONLINE TEST

Test yourself on this topic at www.brightredbooks.net

STRUCTURES: VECTORS AND CONCURRENT SYSTEMS

VECTORS

A vector is a type of force that has both magnitude and direction. Because of this, vectors are usually represented by a line, or 'vector quantity'. The direction of the force is shown by an arrow-headed line, and the length of the line represents the size of force.

EXAMPLE:

A cyclist is pedalling along a road with a force of 700N, but as he is cycling he has an additional force of a tailwind helping him. This has a force of 150N. The friction created by the tyres on the road, though, will hinder him by 200N.

The overall effect will be 700 + 150 – 200. This will give a vector force of 650N.

A suitable scale can then be selected to represent a Newton.
In this case, 1N = 1mm. This is called a 'vector diagram'.

700N + 150N – 200N = 650N

When the 3 forces are added together, they can be replaced by a single force that has the same effect. This is known as the 'resultant'.

Resultant = 650N

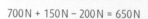

RESULTANT = A + B

RESULTANT = A + B

VECTORS AT ANGLES

It is extremely unusual for forces to be at 180° to each other. Instead, they will be at some sort of angle to each other. By joining the angles head to tail, we can work out the resultant by creating a triangle of forces.

It doesn't matter which direction the forces are joined in, as the resultant will always be the same – they just have to be connected by the head of one arrow to the tail of the other. The resultant will be the direction in which the object will move – and the force of this movement – with both these forces acting on it.

The equilibrium will be the exact same as the resultant, just turned around 180° to ensure that the system would be in perfect balance.

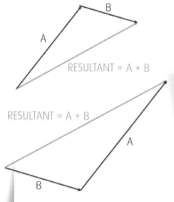

DON'T FORGET

The arrows **always** have to be head to tail as shown, or it will not work!

EXAMPLE:

WIND 200N

ENGINE 800N

WIND 200N

ENGINE 800N

By drawing this to scale (in this case, it is 1N = 1mm), you can measure the resultant to discover the force this vector has. It was 720mm, so it means the resultant force is 720N.

CONCURRENT SYSTEMS

Using this same graphical technique, it is possible to work out the forces within a system.

EXAMPLE:

A semi-submersible oil platform weighing 31 MN is moved into position by 2 tugboats.

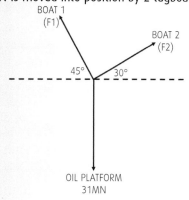

BOAT 1 (F1)

BOAT 2 (F2)

45°   30°

OIL PLATFORM
31MN

ONLINE

Watch the clip at
www.brightredbooks.net
to help with your
understanding of vectors.

Step 1

Draw a line to a suitable scale that represents the known force. In this case, the line will be 31 mm long, with 1 mm = 1 MN.

Step 2

At one end of the line, draw a line at the correct angle to represent one of the unknown forces. This will need to be drawn using a protractor to ensure the correct angle.

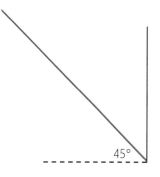

45°

Step 3

At the other end of the line, draw another line at the correct angle to represent the other force.

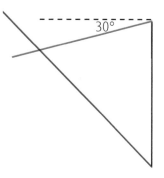

30°

Step 4

Using a ruler, each line can now be measured. By using the same scale we used previously, we can now transfer this into Newtons to find out the forces.

In this case   F1 = 28 MN approx.
                   F2 = 23 MN approx.

 THINGS TO DO AND THINK ABOUT

Part A
45
Part B   Part C

A hanging basket is supported by the frame shown here.

The force in part A is 5000 N, and the force in part C is 2000 N. Determine the size of the force in part B by drawing a triangle-of-forces diagram.

COURSE ASSESSMENT

COURSE ASSIGNMENT

WHAT IS THE ASSIGNMENT?

The course assignment is your opportunity to show your understanding of complex engineering concepts. This assignment is graded out of 50 marks, approximately 31%, of your overall grade, so it is worthwhile putting in the time and effort to do as well as you can.

The assignment is a closed-book assessment, which means you will not get any help from your teacher and you will not be allowed to look at your notes or to use the internet to help you. If a problem occurs and you do not know how to proceed, move on to a different task. You have eight hours to complete the whole assignment and there is no need to do it in the given order.

The assessment task you are given differs every year and will contain instructions and details of any equipment or materials required.

Marks will be awarded for:
- analysing the problem
- designing a solution
- building a solution
- testing the solution
- evaluating on the solution.

When you are given your assignment, there will be several worksheets at the end to complete these tasks. Make sure you use these. If your writing is not neat, a digital version should also be made available to you so you can type directly into it.

ANALYSING THE PROBLEM

To analyse a system, it is usually broken up into three sections: a system diagram, a sub-system diagram and a specification.

When completing a system diagram, make sure you consider **all** possible inputs and outputs to the system. Take care to properly read the design brief to ensure you do this.

The sub-system diagram breaks this system diagram down into all its possible components. This needs to be detailed and complete to get full marks and must show the interaction between different sub-systems. This means you have to consider all parts of the system and all components involved.

If a specification is needed, this should be a bullet-point list of the things the design must and must not do. This information will be found within the brief you are given, so read through this with a fine-tooth comb.

DESIGNING A SOLUTION

This section could also be broken up into three distinct parts, depending on what problem you are given: designing a flowchart, designing any mechanical or electronic sub-systems and designing the structural sub-systems.

If you have to design the flowchart, it is worthwhile annotating it to help show your thinking. You could annotate your design by putting notes next to each different part explaining how it relates to the specification. This could gain you marks for the evaluation section as it shows you are constantly evaluating your solution.

The design of the mechanical aspect of your solution should contain sketches of the gearing system/ mechanical aspects with annotations explaining your solution, as well as any relevant calculations to prove it does what it is expected to do.

Similar to the design of the mechanical aspects, the design of the structural aspects should also contain detailed sketches. This time, they should be of each structural sub-system – and once again, they should be clearly annotated to explain your thought process.

TESTING THE SOLUTION

It is likely that you will be asked to complete a test plan to check if your design complies to the specifications. Your answer should be logical and detailed and fully explain what test you are completing, what is expected to happen, and what actually happens.

CONSTRUCTING/SIMULATING A SOLUTION

This can potentially be broken into four distinct parts: the construction and/or simulation of the mechanical control sub-systems, the justification of materials that you would use, the construction and/or simulation of any electronic and control sub-systems, and the microcontroller code.

Once you have constructed/simulated the design, annotate it. Point out what is what on your design, and **fully** explain it. For example, if your solution involves gearing, take a photo, or print out the simulation, and write on it to point out which gear is which. What type of gearing system have you used? Why? How many teeth does it have? What is its purpose within the system? And so on.

You could even write about your materials, and justify them in a page like this. What material have you used in this aspect of your solution? Why? Compare this to several other materials to ensure you have chosen the correct one, and justify your decision.

When constructing or simulating your electronic aspect of the design, you must ensure that all input and outputs are connected properly to the microcontroller and **fully** simulated. Once again, this design should be fully annotated, explaining the purpose of each component. If you have to replace a component, explain this decision and justify it.

As with the mechanical-design section, the justification of materials could be integrated into this if you wish. State the components you used in your solution and justify your thought process, comparing this to other components and systems to ensure you have chosen the correct one.

If a microcontroller is needed in your solution, a print-out of your microcontroller code should also be given, and it should fully match your flowchart design. If your flowchart has changed, redraw it – just make sure it still matches the brief and your specification.

EVALUATION

The evaluation is a way of testing that your solution fulfils all the needs of the specification. To complete this section, break it down into the different specifications and answer each part one at a time. It is vital you answer this by: stating what the specification is so you know what it is you are evaluating; whether the specification was actually met; how you know it has been met; and what you had to do to ensure that the specification was met.

 THINGS TO DO AND THINK ABOUT

Go through the band descriptors of the marking scheme and try to mark your own work. Make sure the work you hand in covers everything asked.

Also, make sure you read over your assignment and ensure it is clear, in order and well laid out. Your teacher will not give you back your assignment to clear things up, or any feedback to further improve it. Make sure it is the best work you can possibly do before you hand it in.

 DON'T FORGET

When printing out your work, make sure it is clear and no parts have been cropped off. If the marker cannot see your work, they cannot give you the marks!

 DON'T FORGET

If you are asked to construct or simulate a model of something that is already there, copy it **exactly**! Even if you see errors, **do not** change it... at least not yet! Make sure all pins used are the same, orientation of components is the same and yes/no options in decision boxes are the same! It is likely going to ask you to fix any issues **later**.

ONLINE

The assignments from previous years, as well as the answers, can be found on the past paper section of the SQA website.

WRITTEN PAPER

The question paper assesses your knowledge from all three units, giving you the opportunity to show that you have gained a depth of knowledge and understanding and are able to answer appropriately challenging questions.

The exam paper will be broken into two separate sections.

ONLINE

Don't allow yourself to get too stressed. Yes, it is important to study – but plan some short study breaks to take your mind off it for a period of time. Go to www.brightredbooks.net to get some good advice on how to relax and de-stress yourself.

VIDEO LINK

Watch the video at www.brightredbooks.net for advice on creating an effective study plan.

DON'T FORGET

Remember to use your data booklet in the exam. It has most of the calculations and units that you will need within it.

SECTION 1

Section 1 will be out of 20 marks, and will consist of short-answer questions. These are similar to the questions you have encountered throughout this book. By reading this book and completing these questions, along with your work in school, you will have developed and prepared yourself for them. Each question will focus completely on one individual aspect of the course, and they will be designed for you to show the knowledge, understanding and skills that you have gained throughout the National 5 Engineering Science course.

SECTION 2

Section 2 will be out of 90 marks, and will consist of extended-response questions. These will contain numerous parts of the course within the one question, and are designed to show you that many aspects of engineering overlap and how they can be all connected in a real-world situation. This section tests not only your knowledge but also whether you have the understanding to tackle this.

⚙ ACTIVITY:

The following questions are examples of what you might see in Section 2 and have been designed to help you to build up your proficiency in approaching this type of question.

1 A motorised system is used to move a lighting gantry within a theatre.
   (a) Calculate the rotational speed of gear D, if the motor shaft rotates at 750 rev min^{-1}.
   (b State the name of a suitable mechanism that could be used to stop gear D from slipping down the rack.
   (c) When fully loaded, the lighting gantry has a mass of 500 kg.
      Calculate, showing all working and units:
      (i)   The work done moving the lighting gantry by 15 m.
      (ii) The electrical energy supplied to the motor if the system is 82% efficient.
      (iii) The power consumption of the motor in 45 seconds.
   (d) State one method of reducing energy losses when the gantry is being moved.
   (e) The control diagram for the lifting system is shown here.

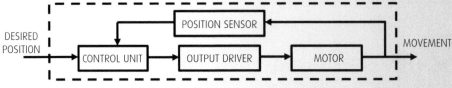

With reference to the diagram, describe the operation of the system when a new position is set.

2 Here is shown the circuit diagram for an automatic water heater.

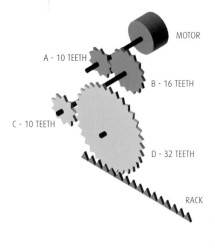

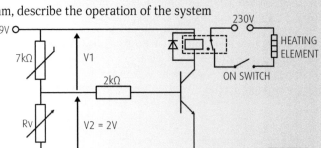

contd

(a) Using appropriate terminology, describe the operation of the circuit when the on switch is pressed to heat the water.

(b) An SPDT relay is used in the system.
  (i) State the full name of this relay.
  (ii) Explain why a relay is required to operate the heating element.
  (iii) State the function of the variable resistor in this circuit.

(c) When the transistor is saturated, V2 is found to be 2 V.
  Calculate:
  (i) The voltage V1.
  (ii) The resistance of the variable resistor RV.
  The 230 V, 12 A heating element takes 15 minutes to heat 20 kg of water to 35°C.

(d) Calculate:
  (i) The electrical energy supplied to the heating element.
  (ii) The heat energy transferred to the water if the system is 75% efficient.
  (iii) The change in temperature of the 20 kg of water during this time.
  (iv) The starting temperature of the water, if the final temperature is 35°C.

(e) Describe two reasons why it is important to conserve energy.

3 Many engineers would be involved in the design and construction of a rollercoaster.

(a) Describe two tasks that a structural engineer would undertake during its design.

(b) An electronics engineer may use computer simulation during the design process. State one feature of the rollercoaster design that the electronics engineer could simulate.

(c) The logic diagram for part of the electronic control system used in the rollercoaster is shown here.
Complete the Boolean equation for the logic diagram.
Z =

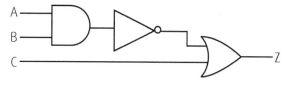

(d) The logic diagram for a second part of the electronic control system is shown here.

Complete the truth table below for the logic diagram.

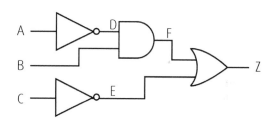

A	B	C	D	E	F	Z
0	0	0				
0	0	1				
0	1	0				
0	1	1				
1	0	0				
1	0	1				
1	1	0				
1	1	1				

(e) The rollercoaster carriages sit on the track as shown here.
The forces acting on the structure are as shown in the free-body diagram.
Calculate the size of reaction force RA, by taking moments about RB.
Show all working and the final unit.

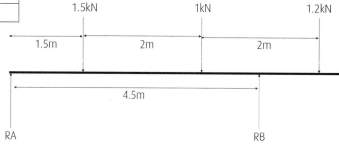

THINGS TO DO AND THINK ABOUT

If you have worked hard throughout the year and you have put in a sufficient level of studying, the exam should be nothing to fear. Your in-class work, and your own self-studying with this book, will have given you a great platform for the exam. You have the knowledge and understanding – now just believe in yourself and your abilities. You can do it!

DON'T FORGET

When you are sitting the exam, ignore everyone else around you and focus on yourself. Concentrate on the questions and your answers, and not on anyone around you who may be scribbling away. Focus and answer your questions in a timely manner – if this means taking a few seconds to re-gather your thoughts or to read the data booklet to find a calculation, then do this!

ALU

Arithmetic and Logic Unit. It operates in a microcontroller by reading instructions from the ROM and then carrying out the mathematical operations for each instruction.

Analogue

A type of electronic signal that can fluctuate between on and off.

Boolean expression

A mathematical equation showing how a logic circuit works.

Bus

Carries information between the various blocks of a microcontroller.

Carbon footprint

The measurement of the total greenhouse gas emissions that are caused directly and indirectly by a person or organisation.

Clock

Controls the speed at which operations occur in a microcontroller.

Closed-loop control

A system where the output is constantly monitored to ensure that the reference value (one that is set by the user) matches the actual value. If there is any difference between them, the system will change things to reduce the output error to zero.

Continuous loop

Part of a computer program that is designed to allow the same part of a program to repeat an infinite amount of times.

Counter

Part of a computer program that is designed to allow the same part of a program to repeat a set number of times.

Current

The rate of flow of electrons going through an electronic circuit. This is measured in amps (A).

Digital

A type of electronic signal that must be either on or off (1 or 0).

EEPROM

Electrically Erasable Programmable Read-Only Memory. This is a special type of memory that runs on a microcontroller. Like normal ROM, it keeps the program when the power supply is removed; but this can be reprogrammed when desired.

Electrical energy

The energy that is caused by moving electric charges. Electrical energy = Voltage × Current × Time.

Electron

A sub-atomic particle of electrical energy.

Energy

Energy is needed to make things work. It comes in many forms and can be converted from one form to another, but it cannot be destroyed.

Energy audit

A systems diagram showing the total energy in and the total energy out, including waste energy. This should also include all relevant figures, such as the amounts of known energies and percentages.

Energy efficiency

The act of trying to use less energy to provide the same amount of power. Efficiency = Useful energy output ÷ Total energy in.

Energy loss

Energy cannot be destroyed, but it can be transferred into an unwanted form. This is usually heat and sound.

Energy transfer

The transfer of one type of energy to another. Some energies are directly inter-transferable, but some need to be changed multiple times until we get the desired output.

Engineer

A qualified person who will work on engineering solutions. Depending on the speciality, it defines what they do. For example, electronics engineers are concerned with the designing of electronic control systems and circuitry.

Flowchart

A graphical representation of a computer language. This is used to break down the program into smaller chunks and make it easier to understand.

Fossil fuels
These are a very concentrated form of energy that was created by the Earth millions of years ago. This includes oil, coal, natural gas and peat. The energy is created through burning these. They create a lot of energy but release a lot of harmful emissions.

Gravitational pull
The force of gravity exerted on an object by the Earth. This is $9.8\,ms^{-2}$, and if you are given a figure measured in kg, multiply it by this to get the weight in Newtons.

Greenhouse effect
When fossil fuels are burned, they release harmful gases that pollute the atmosphere. This stops heat escaping from the Earth.

Heat energy
The energy created by heating up an object. Heat energy = heat capacity of material x mass x change in temperature.

High-level computer language
A complex computer language that can be written to communicate with the microcontroller. Variations of this include C, BASIC and Python.

Impacts of engineering
The effects that an engineering solution will have on the local and surrounding area. These could be social, economic or environmental.

Infrastructure
The basic, underlying framework of an engineering system.

Inputs
What goes into a system to make it work.

Input transducers
Input devices that are used to convert a change of physical conditions into a change in resistance and/or voltage. Examples of these are thermistors, LDRs and variable resistors.

Inverter
Another name for a NOT logic gate.

I/O port
The communication lines between a microcontroller and the 'real world'. This is where any input or output device would be connected.

Kinetic energy
The energy that exists through movement.
Kinetic energy = ½ × mass × velocity².

Labels
Used within high-level-language computer programs to break the program into branches so that it is easier to understand.

Logic gates
Basic building blocks of a digital circuit, used to process a combination of different inputs. These can be AND, OR or NOT gates.

Logic gate: AND
A logic gate where input A AND input B must be on for the output to be on.

Logic gate: NOT
A logic gate where input A is inverted. For example, if the input is logic level 1, the output will be 0.

Logic gate: OR
A logic gate where input A OR input B must be on for the output to be on.

Microcontroller
A small microchip that can be used for sensing inputs from the real world, and then controlling output devices based on this. This can be reprogrammed.

Ohm's Law
The calculation for working out the relationship in a circuit between voltage, current and resistance: voltage = current × resistance.

Open-loop control
A system that has no form of feedback.

Outputs
What comes out of a system.

Parallel circuit
A type of electronic circuit that is broken up into different branches where electrons have a 'choice' where to go.

Potential divider circuits
See 'Voltage-divider circuits'.

Potential energy
Stored energy within an object due to its vertical position or height. The energy is stored as the result of the gravitational attraction of the Earth: potential energy = mass × gravity × height.

Power
The measure of the rate of energy transfer. It is the amount of energy that is produced or absorbed within a circuit. This is measured in watts (W).

power = energy transfer ÷ time

power = current × voltage

RAM
Random-Access Memory. This is a temporary memory that is used for storing information while a computer program is running.

Relay
A type of electromagnetic switch. It is used to join two electrical circuits together, allowing a low-powered circuit to control another one that has a far higher power supply.

Renewable energy
Energy that can be harnessed from natural resources such as wind, water and sun. This type of energy will not run out.

Resistance
The measure of how much voltage is required to let current flow in a circuit. This is measured in ohms (Ω).

ROM
Read-Only Memory. This contains the operating instructions for the microcontroller. The ROM is 'programmed' before the microcontroller is installed, and the memory retains the information even when the power is removed.

Series circuit
A type of electronic circuit that has all components joined up together in one continuous loop for the electrons to flow around.

Sub-procedure
A small section of code that can be called upon by the microcontroller but is not within the main program.

Sub-system diagram
This is an expanded version of a universal system diagram to show the internal features of the system.

System boundary
This can be found within a sub-system diagram. It is a dashed box containing the internal features of a system.

Systems approach
Looking at an engineering solution and breaking it down into its main input and output components.

Transistor
A semiconductor device that uses small electrical currents to control much larger currents. It acts as a digital switch and saturates at 0·7 V.

Truth table
A table made up of binary 1s and 0s that represent the inputs and outputs of a logic circuit.

Universal system diagram
A drawing that shows a system broken down into inputs, process and outputs.

Voltage
Drives the electrons through the components within an electrical circuit. Voltage is measured in volts (V).

Voltage-divider circuits
A circuit where two resistors are connected in series, with the output coming from in between these resistors. One of these resistors is usually an input transducer, and using it in this circuit allows the change in resistance to be converted into a voltage change, allowing the change in physical conditions to be processed.

$$V_{out} = \frac{R_2}{R_1 + R_2} \times V_{cc}$$

$$\frac{V_1}{V_2} = \frac{R_1}{R_2}$$